Mitigation of the Rural Fire Problem

Strategies Based on Original Research and
Adaptation of Existing Best Practices

Mitigation of the Rural Fire Problem

Strategies Based on Original Research and
Adaptation of Existing Best Practices

Final Report of Cooperative Agreement EME-2004-CA-0187

Sharon Gamache
John R. Hall, Jr.
Marty Ahrens
Geri Penney
Ed Kirtley

December 2007

Table of Contents

List of Tables and Figures

Tables

Figures

continued on next page

Executive Summary

In the Spring of 2004, the U.S. Fire Administration (USFA) partnered with the National Fire Protection Association (NFPA) in a cooperative agreement project entitled *Mitigating the Rural Fire Problem.* The purpose of the project was to examine what can be done to reduce the high death rate from fires in rural U.S. communities.

Rural communities, defined by the U.S. Census Bureau as communities with less than 2,500 population, have a fire death rate twice the national average. The objectives of the project were to a) conduct research on behaviors and other factors contributing to the rural fire problem, b) identify mitigation programs, technologies, and strategies to address those problems, and c) propose actions that USFA Public Education Division can take to better implement programs in rural communities.

Research sources included a review of the published literature, some original statistical analysis, and information from national technical experts who have worked with NFPA.

Characteristics of rural America

The primary defining characteristic of rural America is separation—separation of communities from one another and separation of residents from one another. The low density of rural communities means a loss of economies of scale and of concentration. Tasks involving travel take more time and cost more. The potential market for any business requiring travel—either for delivery of the product to the home or for the resident to acquire the product at a store—is smaller, and that affects costs of operation and revenues. Print media are among the businesses so affected, and this has an impact on the quantity and ease of communication within and to a rural community.

The most important correlated characteristic of rural America is a greater likelihood of being poor. For example, in 2003, the percentage of the population below the poverty line was 12.1 percent inside metropolitan areas and 14.2 percent outside metropolitan areas. Less income means fewer resources. It means a greater need for safety—in the form of safer (often newer) products and in devices designed to provide safety (such as smoke alarms)—and a reduced ability to fill that need without outside help.

Poverty is more important than distance as a factor driving the higher fire risk in rural America. In the U.S., the highest-risk communities are the smallest and the largest communities—the rural communities and the largest cities. Rural communities and large cities do not have distance and separation in common, but they do both have a higher likelihood of poverty.

1

Other important characteristics of rural America have to do with the social networks that organize life and the importance of trust and familiarity to the operation of these networks. It is not clear whether these conditions are very different in larger communities, but our experts on fires and fire safety in rural communities all agreed about the importance of these networks in rural life.

Characteristics of the rural fire problem

The distribution of incident types is roughly the same in rural and nonrural areas, including the proportion of reported outdoor fires involving neither structure nor vehicle. However, the cause profiles are quite different for both outdoor fires and residential structure fires.

Forty-five percent of the rural outside fires were caused by open flame, 16 percent by arson, and nine percent by natural causes. By contrast, arson caused 44 percent of the nonrural outside fires.

Rural residential fires were more likely to be caused by heating equipment, to occur in properties that had no smoke alarms at all, and to have flame damage extend to the entire structure. Thirty-six percent of the rural residential fires were caused by heating, 13 percent by cooking, and 12 percent by electrical distribution equipment. Twenty-six percent of the fatal residential rural fires were caused by heating, 23 percent by smoking, and 17 percent by electrical distribution equipment. Smoking caused 28 percent of the nonrural fatal residential fires; 17 percent were arson, and heating caused 12 percent.

Fixed area heaters, including wood stoves, were involved in 38 percent of the rural residential heating fires. Chimneys (25 percent) ranked second and fireplaces (11 percent) ranked third. Adhesive, resin, or tar was the type of material first ignited in nearly half of the rural residential heating fires. Sawn wood was first ignited in 19 percent of these fires.

Almost three-quarters (73 percent) of rural residential fires occurred in properties without working smoke alarms, compared to 65 percent in nonrural properties. The larger difference seen was in presence of smoke alarms versus nonoperating smoke alarms. In 58 percent of the rural residential fires, no smoke alarms were present at all; in 15 percent, smoke alarms were present but not operating. In 42 percent of the nonrural incidents, no smoke alarms were present at all; in 23 percent, these devices were present but not operating.

Flame damage extended to the entire structure in 29 percent of the rural residential structure fires but only 17 percent of such incidents in nonrural areas.

Rural differences by region

Rural communities and the rural fire problem differ considerably from one U.S. region to another. From 2000 to 2004, the rural fire death rates per million population were 29.0 for the South, 28.2 for the West, 27.0 for the Northeast, and 22.8 for the north-central region. Historically, the South has had the highest fire death rate and the highest rural fire death rate, but in recent years, the gap between the South and the other regions has shrunk, and the South has not had the highest fire death rate or the highest rural fire death rate in every single year.

The South (sometimes called the Southeast) is also by far the most populous region (more than one-third of total U.S. population) and contains nearly half the total U.S. rural population. It is not uncommon, therefore, to talk about the rural fire problem and the South's fire problem interchangeably. This is misleading. Rural communities have the highest fire incident and fire death rates in every one of the four regions in most years.

Because the South has half the Nation's rural population and has a higher rural fire incident rate and rural fire death rate than other regions, the South dominates total national rural fire statistics. For example, the heightened share of rural fire deaths involving heating equipment is as much a South phenomenon as it is a rural phenomenon. Of the four regions, the South has the most consistently mild and short heating season. Therefore, poorer households in the South are the ones who find it most feasible to try to use space heating exclusively, resulting in the fire experience repeatedly documented for space heating as compared with central heating.

Often, fire risk is correlated with poverty, and rural areas tend to be poorer than nonrural areas anywhere in the country. However, the gap in poverty rates between rural and nonrural areas is largest in the South and is associated more with African-Americans than is true in other regions. The rural South has nearly all of America's rural African-Americans and nearly all of America's poor rural African-Americans. Rural populations tend to have a lower African-American share than do nonrural populations, but that difference is almost nonexistent in the South. The rural poor do tend to have a lower African-American share than do the nonrural poor, but that difference is much more pronounced outside the South.

The West historically has the lowest overall fire death rates but not always the lowest rural fire death rates. This region contains some distinctive and important subgroups of the rural population that deserve separate attention, but also should not be mistaken for the typical rural population of those regions. The West includes Native American communities, migrant worker communities, and Mexican border communities, sometimes called "colonias." As with the African-American population of the rural South, so with the Native American and Mexican-American populations of the West; each region's poor rural populations have a distinctive character.

It is also a mistake to think of rural America as primarily a farming community. Nonfarm rural dwellers outnumber rural farmers by about 18-to-1. However, this imbalance varies by region. The north-central region still has much of America's agricultural activity and farms.

In addition, poor housing quality generally is a problem in the rural South. The South also has the highest proportion of housing units in manufactured homes—12 percent versus 3 to 7 percent in the other regions. Historically, this has been a factor in the elevated fire death rate in the South, because until very recently manufactured homes have had a higher fire death rate than conventional "stick-built" homes or apartments. However, now that most manufactured homes in use were built after the advent of the construction requirements of the U.S. Department of Housing and Urban Development (HUD), introduced in 1976, manufactured homes are no longer a high-risk environment. This may be part of the reason why fire death rates in the South are no longer consistently much higher than rates in other regions.

The rural fire service

The principal distinguishing characteristics of the rural fire service are

(a) Nearly all (99.5 percent) of the rural fire departments are all- or mostly volunteer.

(b) Rural fire departments are more likely to have insufficient companies and personnel to meet national guidelines for effective response, as travel distances and travel times to fires and other emergencies tend to be longer because of the low density of such communities.

(c) Rural fire departments are less likely to have needed equipment.

(d) Rural firefighters are less likely to have needed training.

(e) Rural fire departments are less likely to conduct fire prevention programs of all types, including code enforcement.

Guidance for effective rural safety programs

This section summarizes information from Chapter 7. In that chapter we provide guidance to volunteer firefighters, health and safety organizations serving rural communities, and other community leaders on how best to implement programs in rural areas.

Rural communities have special challenges in program distribution. Programs that operate through local codes and regulations need to address gaps in code enforcement, which tend to be greater in rural communities. Programs that are delivered via mass media need to be realistic about usage and access rates in rural communities for the selected mass media and the target population for the programs. Mass media may be less established in rural communities making it less effective for use in fire safety programs. Under these conditions, door-to-door distribution is the more effective option. Programs that are delivered in person need to address the lower densities and greater place-to-place distances in rural communities. Pay particular attention to the program's ability to reach all or nearly all of its target audience.

A network, as the term is used here, consists of a central organization with existing relationships for particular purposes with a larger group of people in the community. Several existing networks common to rural areas have been identified as potentially valuable to and supportive of fire safety programs. Build around existing networks. Build a relationship of trust between those delivering a program and those targeted to receive the program. Try to work with a rural community's key influential people.

The central organizations for these identified existing networks are as follows:

- fire departments (nearly all volunteer in rural communities);

- health care (including both public health personnel and the individual private care providers, who may be the only ones located in a rural community);

- churches and other faith groups;

- schools;

- area agencies on aging, senior citizen centers, Meals on Wheels (people age 65 and older are a high-risk group for fire death);

- rural electrical cooperatives;

- Fire Corps;

- national nonprofit and service organizations, American Red Cross, and Safe Kids; and

- cooperative extension.

Consult available books and other resources on best practices in the management of nonprofit organizations, with particular attention to the oversight and effective use of volunteers. Plan for delays and setbacks, and be ready to adapt or respond as needed. Consult the literature on rural safety and health program design and delivery.

Modify model programs to reflect local conditions and provide local relevance. Make sure, however, that the information used to identify local conditions is current and fact-based, and that the changes do not damage the program elements that are essential to its design and effectiveness.

Recommendations

The following are recommendations for further research and for implementation of programs for USFA, NFPA, and other national and local organizations interested in mitigating the rural fire problem.

1. We recommend the development and implementation of a model multihazard survey for homes that could be incorporated as a voluntary outreach program and used to identify homes that need changes in their equipment. Equipment checked could include both portable and stationary space heaters, electrical wiring and related parts of the electrical distribution system, and smoke alarms. The survey also could check related conditions, such as locked, blocked, or inoperable doors and windows that are part of primary or alternate escape routes.

For greatest effectiveness and least burden on the households, the survey would be conducted by trained professionals, though not necessarily certified fire inspectors or electricians, with the consent of the households. Despite the term "survey," this is not envisioned as a hand-off instrument for households to use to review their own equipment.

After the survey the residents of the household would be given a list of prioritized safety hazards that should be corrected. In an ideal program there would be community block grants or other funding that would help the property owner to follow through on some of the improvements suggested by the survey.

2. As an enhancement to recommendation #1, we further recommend the production of a walk-through video showing a home survey in a rural home.

3. We recommend partnering with national and regional organizations and agencies such as the U.S. Environment Protection Agency (EPA), the Southwest Indian Foundation; the U.S. Department of Agriculture (DOA); the Hearth, Patio and Barbecue Association; and the HEARTH Education Foundation to develop programs that would replace problem space heaters.

4. We recommend the development of a program for improvement of rural electrical system safety that will set priorities in terms of the range of hazards and conditions that may be identified in a survey and will identify affordable modifications suitable for use in existing homes.

5. We recommend the development of a national strategy to install working smoke alarms in every rural home.

6. We recommend the development of a DVD/video that would communicate the importance of reaching rural communities and would portray the variety of rural communities in the United States by region and group type.

7. We recommend development of organizational options for providing a supportive network that could be extended to every rural community.

8. We recommend increased research on effective ways to meet the needs of the rural fire service.

Chapter 1. *Introduction and Project History*

In the Spring of 2004 the U.S. Fire Administration (USFA) partnered with the National Fire Protection Association (NFPA) in a cooperative agreement project entitled *Mitigating the Rural Fire Problem*. The purpose of the project was to examine what can be done to reduce the high death rate from fires in U.S. rural communities.

Rural communities, defined as communities with less than 2,500 population, have a fire death rate twice the national average. The objectives of the project were to a) conduct research on behaviors and other factors contributing to the rural fire problem, b) identify mitigation programs, technologies, and strategies to address those problems, and c) propose actions that the USFA Public Education Division can take to better implement programs in rural communities.

Prior to this project, two reports on the rural fire problem had been issued by the USFA: *The Rural Fire Problem in the United States* (FA-180, August 1997) and *A Profile of the Rural Fire Problem in the United States* (FA-181, August 1998), the latter being a brief summary of the former. These reports provided the starting point for our understanding of the size and characteristics of the rural fire problem. Among the key findings of these earlier reports were that the three leading causes of fire deaths were heating equipment, smoking, and electrical distribution and lighting equipment and that rural areas have a lower percentage of smoke alarms present in residential fires than do non-rural communities.

Work plan

In May of 2004, NFPA met with Bill Troup, USFA COTR (contracting officer's technical representative) for the cooperative agreement, to discuss and reach agreement on the proposed work plan for the grant. Phone meetings with USFA have been held periodically throughout the project. Quarterly update reports have been submitted throughout the project duration.

Literature review

NFPA conducted a literature review of studies of the rural fire problem and of keys to success in rural fire safety programs. Studies and articles describing other successful programs implemented in rural America on other issues were also reviewed.

NFPA also provided original statistical analysis to add to the findings of the two USFA reports and provide further insights into the size and characteristics of the rural fire problem.

See Appendix A for the complete deliverable from this literature review and statistical analysis. See Appendix B for the material specifically on fire service needs and firefighter health and safety.

Information gathering from national and local organizations

After the completion of the literature review, information was obtained information from national level technical experts who have worked with NFPA.

National groups interviewed regarding fire safety programs in rural communities were: International Association of Fire Fighters (IAFF), International Association of Fire Chiefs (IAFC), National Volunteer Fire Council (NVFC), National Association of State Fire Marshals (NASFM), International Fire Marshals Association (IFMA), Safe Kids World-wide, and the Centers for Disease Control and Prevention (CDC).

Research also was done into organizations and agencies that potentially had important rural fire safety programs on specific issues, even if those programs were a relatively small part of their overall activity. Those included programs by such groups as the Chimney Safety Institute of America, the Environmental Protection Agency (EPA), and the National Chimney Sweep Guild.

The literature review, statistical analysis, and information from experts were intended to add to the project's understanding of key risk factors. This included looking for evidence to support or refute hypotheses about the risk impact of characteristics of the rural population (e.g., poverty), their homes (e.g., manufactured homes, older homes), and the products in their homes (e.g., certain types of heaters, older electrical systems). Similarly, research on programs extended to behavioral aspects (i.e., fire safety education and supporting behavioral research), technological aspects (e.g., smoke alarms, home fire sprinklers, building construction, characteristics of electrical systems), enforcement aspects (e.g., codes and standards), and fire department response aspects (e.g., water supply availability, training, and equipment for firefighters). Comprehensive, well-designed, thoroughly evaluated programs, such as the 14 CDC-funded 5-year State programs that implement smoke alarm installation and fire safety education projects in high-risk communities, were studied in great detail.

In addition to organizations and agencies that already were involved in the delivery of rural fire safety programs, the project looked for other agencies and organizations that had an established interest and focus on rural safety or quality of life and that had established networks in rural areas, which could form the basis for fire safety program delivery networks. This led to discussions with representatives of the Electrical Safety Foundation International (ESFI), National Rural Electric Cooperative Associations (NRECA), and the National Rural Health Association (NRHA).

Native American communities and so-called colonias were recognized early as distinctive types of rural communities that deserved individual attention. See Appendix C for a complete list of people who were consulted by phone or in person.

Meetings at NFPA

After initial information was gathered, NFPA hosted two separate 2-day meetings at its headquarters. The first group, which met on October 24 and 25, 2005, were representatives of "outreach" organizations, that demonstrated particularly strong delivery methods and networks for whatever programs they implemented in rural areas. This group was invited to identify successful model programs with associated networking systems that could be adapted to the rural fire problem. Representatives attended from the following organizations:

- Arizona Indian Health Service Injury Prevention Program on the Navajo Nation;

- Electrical Safety Foundation Institute;

- American Burn Association;

- National Rural Electric Cooperative;

- Safe Kids Worldwide;

- Louisiana State University;

- National Rural Health Association; and

- South Carolina Department of Health and Environmental Control, (CDC smoke alarm installation program).

A second meeting was held on October 27 and 28, 2005, with representatives of national and local fire service organizations that have included, as part of their mission, a focus on the fire problem in rural areas. Representatives were invited to discuss their experiences working in rural areas or their experience with issues related to fires and burns in rural areas.

Representatives from the following groups participated in these meetings:

- International Fire Marshals Association;

- Georgia Department of Public Health, CDC Smoke Alarm installation program;

- Native American/Alaska Native Fire Chiefs Association;

- International Association of Fire Chiefs—(Fire Life Safety Section Representative and Volunteer Combination Officers Section Representatives);

- National Volunteer Fire Council;

- Centers for Disease Control and Prevention/National Center for Injury Prevention and Control/Division of Unintentional Injury Prevention;

- National Association of State Fire Marshals; and

- Holmes County, Mississippi, Smoke Alarm Program, (NFPA/USFA funded program).

Gap analysis and best practices

NFPA took the information gathered from the literature review, the statistical analysis, the phone, and in-person individual meetings, and the two (2-day) group meetings at NFPA, and developed a gap analysis. The purpose of the gap analysis was to determine what was needed for success in rural fire safety programs and compare those needs to existing program models and current practices in the field. Unmet needs constituted a gap in rural fire safety. See Appendix D for the full report on the gap analysis.

NFPA developed the "Best Practices for Programs Targeting the Rural Fire Problem." This report provides guidance on best program practices for local volunteer firefighters and other leaders in rural communities, based on the needs for success identified in the gap analysis. See Appendix E for the full report on best practices.

Part of the gap analysis involved additional analysis to flesh out differences in the rural fire problem and related local characteristics of people, homes, and equipment by region or other distinct type of regional community. The importance of these distinctions was one of the major findings of the project.

Key messages

The project team developed key fire safety educational messages that could be the center of every program. (See Chapter 7) The team emphasized messages that would address the major causes of fire deaths in rural areas, other major causes of rural fires, and other key conditions and practices related to rural fire deaths (e.g., usage or non-usage of smoke alarms). The content of the messages was taken from established fire safety education sources. The content usually was not altered because of the rural focus, but the choices of which messages to emphasize was driven by the rural focus.

This task completed the work of identifying most promising rural fire safety strategies. At this point, the project findings included best practices on program delivery and implementation, best messages to use for the educational and awareness components of a program, and best products (e.g., heating equipment, electrical distribution and lighting equipment, smoke alarms) to focus on in changing to a safer environment.

Program guide and PowerPoint® presentation

Efficiencies in the early stages of the project meant that there were sufficient funds to add two deliverables that would format the project findings into a PowerPoint® presentation on project conclusions for even greater ease of use with primary emphasis on the key messages, and a program guide. These were designed for primary use in training volunteer firefighters and other leaders from organizations in rural areas in leading and participating in successful rural fire safety programs, along with secondary use by those program leaders in educating rural residents.

See Appendix F for the PowerPoint® presentation, and Appendix G for the program guide for the rural fire service. This is adapted from the USFA five-step planning guide for use as a short, easy-to-use guide incorporating sidebars on the rural fire project and examples of successful rural fire safety programs.

Development of recommendations

The project team also developed recommendations for further research and for implementation of programs by USFA, NFPA, and other national, State, and local organizations interested in mitigating the rural fire problem.

Review

All material was reviewed in April 30, 2007, by the USFA staff, representatives of major fire organizations, and the participants in the 2-day meetings in October 2005. This review was conducted before completion of the program guide and PowerPoint® presentation, which meant only drafts of material for these two deliverables were included in the review package.

Summary of this report

This final report of the project on mitigation of the rural fire problem has its main text organized around major topics. The appendices represent documentation of major tasks and include more details on the material covered in the text.

Chapter 1 provides an introduction to the project and its background. It also gives a history of the project, including the approaches used in the research and the sources of information that support the project's findings. It describes the components of the project.

Chapter 2 gives an overview of the characteristics of rural America, while Chapter 3 provides an overview of the rural fire problem. There are considerable differences in characteristics among the rural areas of America's four regions—the Northeast, north central (or Midwest), South (or Southeast), and West—and Chapter 4 describes those differences.

Chapter 5 provides an overview of the rural fire service with regard to differences in resources and services provided, particularly differences in response to fires.

Chapter 6 gives the findings on the networks and organizations of rural America where they are relevant to potential programs to mitigate the rural fire problem. Chapter 7 provides guidance for effective rural safety programs. Chapter 8 has recommendations for further research and program implementation.

Appendix A provides the details of the literature review, including original statistical analysis of the rural fire problem.

Appendix B gives rural highlights from the first U.S. fire service needs assessment project, conducted in 2001 and 2002 by NFPA with sponsorship and direction by the USFA. This appendix provides the basis for the findings in Chapter 6.

Appendix B also includes analysis of the number of volunteer firefighters killed on duty and patterns in the circumstances of their deaths. Rural communities cannot be separated out in the firefighter fatality database.

Appendix C has a list of experts who were interviewed in groups or individually, in person, or by phone.

Appendix D provides the project's gap analysis, which was the distillation of the material from Appendix A and the experts listed in Appendix C into more focused Statements of needs, problems, and solutions for the rural fire problem. Appendix D contains the direct basis for the findings in Chapters 3 to 5 and 7.

Appendix E gives the project's compilation of best practices, further distilled from the gap analysis. Appendix E is the primary direct basis for the findings in Chapter 8.

Other deliverables of the project, available with this report from the USFA, are a PowerPoint® slide presentation on the project, suitable for use primarily to train the trainers and other project managers in a rural fire safety program, and a shortened, simplified, and customized version of the USFA's Five-Step Guide to public fire safety education programs, modified for rural needs and circumstances and simplified for a wider target audience.

Chapter 2. *Characteristics of Rural America*

The primary defining characteristic of rural America is separation—separation of communities from one another and separation of residents from one another. This characteristic is inherent in what defines a rural community, which is a community size of less than 2,500 population.

The low density of rural communities means a loss of economies of scale and of concentration. Tasks involving travel take more time and cost more. The potential market for any business requiring travel—either for delivery of the product to the home or for the resident to acquire the product at a store—is smaller, and that affects costs of operation and revenues. Print media are among the businesses so affected, and this has an impact on the quantity and ease of communication within and to a rural community.

The most important correlated characteristic of rural America is a greater likelihood of being poor. For example, in 2003, the percentage of the population below the poverty line was 12.1 percent inside metropolitan areas and 14.2 percent outside metropolitan areas. [1] Less income means fewer resources. It means a greater need for safety—in the form of safer (often newer) products and in devices designed to provide safety (such as smoke alarms)—and a reduced ability to fill that need without outside help.

Poverty is more important than distance as a factor driving the higher fire risk in rural America. The highest-risk community sizes are the smallest and the largest communities—the rural communities and the largest cities. They do not have distance and separation in common, but they do both have a higher likelihood of poverty. The size and characteristics of the rural fire problem is discussed in Chapter 4. Regional differences in rural characteristics and the rural fire problem are discussed in Chapter 5.

Other important characteristics of rural America have to do with the social networks that organize life and the importance of trust and familiarity to the operation of these networks. It is not clear whether these conditions are very different in larger communities, but our experts on fires and fire safety in rural communities all were agreed about the importance of these networks in rural life. One such network—the rural fire service—is discussed in Chapter 6, and an overview of all such networks is provided in Chapter 7.

Distance, remoteness, and isolation

Rural areas are areas with a low density of population per unit area (square mile, square kilometer). The U.S. Census Bureau defines "rural" as a community with less than 2,500 population. Other definitions use different thresholds and may include density criteria and/or criteria based on proximity to cities or other population concentrations.

Many statistics are defined by metropolitan versus nonmetropolitan rather than non-rural versus rural. By either definition, rural communities account for roughly one-fifth of the U.S. population.

The U.S. Census Bureau considers areas other than urbanized areas or urban clusters to be rural. An urbanized area has a nucleus (may or not be a unique city) with at least 50,000 residents. Such an area also has a core of at least one contiguous block group of less than 2 square miles with 1,000 people per square mile. Urban clusters have similar cores, but they have populations of from 2,500 to 49,999.

Metropolitan statistical areas, as defined by the Office of Management and Budget, include "central or core counties with one or more urbanized areas, and outlying counties that are economically tied to the core counties as measured by work commuting." Micropolitan statistical areas include a) nonmetropolitan counties with at least one urban cluster of 10,000 or more residents, and b) noncore counties that lack these urban clusters. Both types of nonmetropolitan counties often are included in studies of rural conditions. [2]

In recent years, the metropolitan statistical areas of the U.S. have typically had 80 percent of the U.S. population in 20 percent of the area. This means metropolitan areas have a population density 16 times the density of nonmetropolitan areas. That, in turn, means the average distance between two points is four times higher in a nonmetropolitan area than in a metropolitan area. In addition, rural areas are the least dense areas in nonmetropolitan areas. Therefore, the distance ratio for rural versus nonrural may be even higher.

In a door-to-door program, quadrupling distance means more time and more cost per household. Consequently, it may result in driving from place to place rather than walking from place to place.

Rural poverty: Characteristics of rural housing

Cushing Dolbeare included housing in areas specifically classified as rural and "other urban" in discussions of rural housing. [3] Rural householders tended to be older, poorer, and more likely to be married and white than their urban counterparts. Housing in rural areas tends to be larger and less expensive than urban housing. Rural renters face a higher housing cost burden than homeowners.

Inadequate housing is a bigger problem in rural households (7 percent) and homes in central cities (8 percent) than in the suburbs (4 percent). [4] Five percent of rural African-American households, 3 percent of rural Hispanic households, and 2 percent of rural white households lived in severely inadequate housing. Seventeen percent of rural African-American households, 10 percent of rural Hispanic households and 4 percent of rural white households lived in housing that was considered moderately inadequate.

Compared to older adults elsewhere in the country, homes of older adults in non-metropolitan areas were more likely to be owner-occupied and to have problems. The nonmetropolitan older adults were more likely to live in manufactured housing than their metropolitan counterparts and to have fewer economic resources. In 1995, more than one million nonmetropolitan housing units occupied by older adults lacked adequate heating equipment. Almost half a million (468,000) nonmetropolitan elder-occupied units had severe or moderate problems with heating, plumbing, electrical systems, maintenance, kitchens, and/or hallways.

Rental housing is less available in nonmetropolitan and rural areas, which can be a challenge for individuals who no longer can cope with the maintenance issues associated with home ownership. [4]

Mass media communications

Rural areas have fewer media providing regular coverage of their local news. This may mean fewer established channels for public awareness campaigns for fire safety. However, it is a myth that rural areas have less Internet access than urban areas. A 2003 survey by the U.S. Department of Commerce (DOC) found that 54.1 percent of rural households had Internet access (61.9 percent had computers) compared to 54.8 percent of urban households (61.7 percent had computers) and 49.3 percent of central city households (56.9 percent had computers). [5] There is no significant difference between rural and urban Internet access.

The same DOC survey found that among the poorest households (under $5,000 per year), the poor rural households had markedly less Internet access (20.0 percent) than the poor urban households (28.4 percent) and less even than the poor central city households (24.3 percent). Therefore, it is true that **poor** rural households, which constitute one of the highest-risk populations in the country, have less Internet access than any other population group defined by economic condition and community size.

Also, rural fire departments are far less likely to have Internet access than are fire departments in larger communities. Only 41 percent of rural communities had Internet access for fire departments compared to at least 63 percent for fire departments in any larger community and at least 93 percent for communities of 25,000 population or more.

Comparable measures of rural versus nonrural access to other forms of mass media have proven elusive. It seems clear, however, that rural choices are often fewer, such as fewer radio and television stations within broadcast range and fewer and smaller newspapers to cover local events. In other words, mass media are less well-established in rural communities, which makes mass media less effective and less attractive for use in fire safety programs. Under these conditions, door-to-door distribution is the more effective option.

Trust and social networks

There is a consensus perception among experts in fire and fire safety in rural America that rural households are slower to give trust to people. Similarity and familiarity are considered key characteristics in gaining trust. Similarity may involve ethnicity, religion, age, culture, region, and language. Familiarity means a lower threshold for trust if some-one unknown is known to a third party who is known and trusted.

It is not clear whether this challenge is distinctive to rural areas. Some very successful fire safety programs in major cities have encountered trust issues in their target popula-tions and solved them by working with and through faith networks.

It was suggested in some of the discussions with experts that urban dwellers are more accustomed to dealing with strangers, suggesting a higher degree of mobility and turn-over in such communities. A U.S. Census Bureau analysis of movement from 1995 to 2000 found that 59 percent of nonmetropolitan dwellers were in the same residence in both years, compared to 53 percent of metropolitan dwellers (and 49 percent of central city dwellers). The survey also found that 80 percent of nonmetropolitan dwellers were in the same county in both years, compared to 79 percent of metropolitan dwellers (and 79 percent of central city dwellers).[6] This supports the notion of a somewhat less mobile rural population, but the difference is not overwhelming, and it is virtually nonexistent at the same-county level.

Another version of this point would be that the lower density of rural areas means a rural dweller is surrounded by far fewer people within his/her range of potential contact and interest. Under these circumstances, it is almost inevitable that urban dwellers will have a much higher fraction of their encounters with people they do not know outside specified roles such as merchant and customer. This would lead back to the same conclu-sion, namely that urban dwellers do not have the option of dealing only with people they know well and, therefore, trust. Rural dwellers do have the option, and the perception is that many of them use it.

There is anecdotal evidence that many people who grew up in rural areas and later went to "the big city" as adults are returning to the rural communities of their youth, possibly to tend to aging parents or to recapture the familiar and comfortable feelings of their youth. These returnees would be more experienced in dealing with strangers, and might be seen as strangers themselves. Furthermore, any vital rural community will have some growth through in-migration, and some of those "transplants" will bring percep-tions and habits from larger communities. At the same time, however, there continues to be anecdotal evidence of rural communities perceived as so lacking in opportunities that every young person who can leave does so.

With so many thousand rural communities, it would not be surprising if all these phe-nomena are true in some rural places. It has not been possible to identify hard evidence of just how much truth there still is in the idea of a generic, insular rural community with sharply distinguished insiders and outsiders.

To the extent that there is some truth to these characterizations, there are implica-tions for fire safety programs. A program without a trusted local advocate is unlikely to succeed. Once the basis of trust has been established with some, however, there may be a very rapid spread of trust to the entire community. There is a lower "tipping point" in a rural community.

References

1. *Income, Poverty, and Health Insurance Coverage in the United States: 2003, http:// www.census.gov/prod/2004pubs/p60-226.pdf* Table 3.

2. Rural Assistance Center, "What is Rural? Frequently Asked Questions," *http:// www.raconline.org/info_guides/ruraldef/ruraldeffaq.php#definition*

3. Dolbeare, Cushing N. "Conditions and Trends in Rural Housing." *Housing in Rural America*, Joseph N. Belden & Robert J. Wiener, eds. California: Sage Publications, 1999, pp. 13-26.

4. Belden, Joseph N. "Housing for the Rural Elderly." Joseph N. Belden & Robert J. Wiener, eds. California: Sage Publications, 1999, pp. 91-97.

5. *A Nation Online 2004, http://www.ntia.doc.gov/ntiahome/dn/index.html*

6. *Migration and Geographic Mobility in Metropolitan and Nonmetropolitan America: 1995 to 2000, http://www.census.gov/prod/2003pubs/censr-9.pdf* Table 1.

Chapter 3. *The Rural Fire Problem—*
Size and Characteristics

Ever since the advent of the NFPA fire experience survey in the late 1970's, it has been possible to compare fire incident and fire death rates relative to population among different sized communities. Rural communities consistently display the highest rates within what is sometimes called a "bathtub" curve, showing high rates at both ends—the smallest and largest communities—and lower rates in the middle.

The rural fire problem also is distinguished by a significantly different cause profile. For the Nation as a whole, the leading causes for home fire deaths (where most fire deaths occur) is smoking materials (i.e., lighted tobacco products) far ahead of intentional fires (e.g., arson), which is far ahead of heating equipment in third place. For rural communities, heating equipment is the leading cause of home fire deaths, followed by smoking materials and, in third place, electrical distribution and lighting equipment.

Outdoor fires account for most fire incidents, though relatively few fire deaths, and they account for roughly equal shares of rural and nonrural reported fires. However, their cause profiles are quite different. In rural areas, unintentional ignitions involving open flame (e.g., open burning, campfires, discarded matches) dominate the cause profile of outdoor fires, while in nonrural areas, intentional fires (e.g., arson) dominate the cause profile.

Rural communities have the highest fire death rates.

During the five-year period 1997 to 2001 (excluding the events of September 11, 2001), rural communities with populations under 2,500 had an average fire death rate of 30.9 per million population. This rate was at least twice that found in most other population intervals with the exception of communities with populations of 2,500 to 5,000, which had a rate of 18.4 fire deaths per million population. Communities with populations under 2,500 averaged 12.0 reported fires per 1,000 population, twice that of all population intervals except 2,500 to 5,000 (8.3), and 5,000 to 10,000 (6.9). These communities also have the highest per capita rate of reported fires. [1]

According to U.S. Census figures assembled by the Northeast Midwest Institute, 59 million, or 21.0 percent, of the U.S. population in 2000 lived in rural areas. Three million, or 1.1 percent, of the U.S. population lived on farms. Fifty-six million, or 19.9 percent, lived in nonfarm rural areas. [2]

Table 1 shows the percentage of rural, farm, and nonfarm rural populations and rank by rural percentage for each State in 2000. These data are combined with data from a table in NFPA's 2004 report on State fire death rates, showing State fire death rates and rank for the 5-year period 1997 to 2001, along with explanatory characteristics and their ranks. [3]

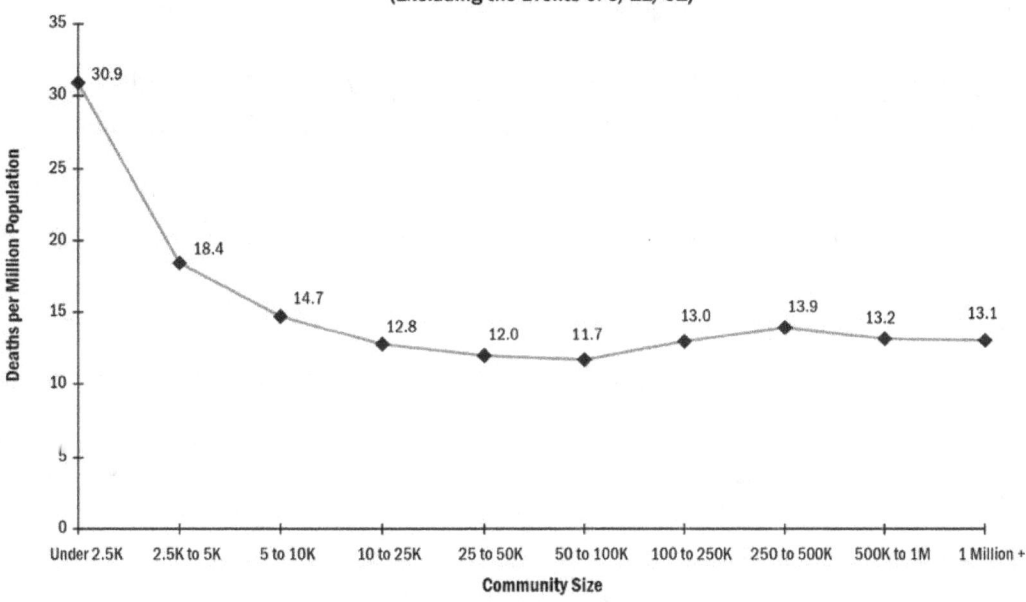

**Civilian Fire Deaths per Million Population by Size of Community:
1997-2001 Annual Averages**
(Excluding the Events of 9/11/01)

Source: *Fire Loss in the United States,* by Michael J. Karter, Jr.

Vermont and Maine ranked first and second in percent of rural populations, but their fire death rates were 12th and 20th, respectively. Among the 15 States with the largest percentages of rural populations, 9 were among the 15 States with the highest fire death rates.

The percentage of rural population is a powerful statistical predictor of fire death rates. NFPA used simple linear regression analyses to estimate the strength (in percent of variation explained) of candidate variables. Lack of education, defined as the percent lacking 12 years of school, accounted for 29 percent of the variation, and was the strongest of the three predictors discussed. The percent of current smokers accounted for 22 percent of the variation, and the percent below the poverty line accounted for 16 percent. Using the same techniques, percent rural explained 31 percent of the variation.

Most high fire death rate States are in the South and have large percentages of rural populations. However, Alaska is the northernmost part of the U.S., and its fire death rate has consistently been high.

The National Fire Protection Association found that almost all States have shown drops in fire deaths and fire death rates over the past two decades. The Southeastern States of the old Confederacy (excluding Florida), plus Alaska, have consistently had fire death rates above the national average. Border States such as Missouri and Oklahoma also tend to have high rates. The study noted that States with small populations can have unusually high death rates in some years due to an increase of just a few deaths. [3]

The rural population in the U.S is not homogenous. Before attempting to address the rural fire problem, it is necessary to understand something about the people, lives, and regional differences of these rural populations.

Table 1. Percentage of Rural Population, Fire Death Rates, and Other Explanatory Characteristics, by State

State	Rural Population in 2000				Fire Deaths per Million Population 1997-2001		Education — Adults without 12 Years of School Average of 1998 and 2000		Smoking — Adults Who Are Current Smokers 1999 percentage		Poverty — People below Poverty Line, Average of 1998 to 2000	
	Rural Population Percent	Percent Rural Population Rank	Farm Dwellers Percent	Nonfarm Rural Percent	Average	Rank	Percent	Rank	Percent	Rank	Percent	Rank
Vermont	61.8 %	1	1.8 %	60.0 %	16.2	12	11.7 %	39	21.8 %	36	10.3 %	28
Maine	59.8 %	2	0.9 %	58.9 %	12.3	20	12.0 %	38	23.3 %	22A	9.8 %	34A
West Virginia	53.9 %	3	1.2 %	52.7 %	18.1	8	23.3 %	1	27.1 %	6B	15.8 %	4B
Mississippi	51.2 %	4	1.6 %	49.6 %	32.1	1	21.2 %	6	23.0 %	25	15.5 %	6
South Dakota	48.1 %	5	7.7 %	40.4 %	9.1	33	11.0 %	42B	22.5 %	27C	9.4 %	38
Arkansas	47.6 %	6	1.9 %	45.6 %	24.2	3	20.8 %	7	27.2 %	4B	15.8 %	4A
Montana	46.0 %	7	4.4 %	41.5 %	10.2	28	10.7 %	44	20.2 %	44	16.0 %	3
Alabama	44.6 %	8	1.2 %	43.3 %	25.9	2	21.9 %	2	23.5 %	20A	14.7 %	8A
Kentucky	44.3 %	9	3.2 %	41.1 %	17.1	10	21.7 %	3	29.7 %	2	12.5 %	18B
North Dakota	44.2 %	10	6.8 %	37.4 %	8.4	38B	15.1 %	23A	22.2 %	34	12.8 %	17
New Hampshire	40.8 %	11	0.4 %	40.4 %	7.9	44B	14.0 %	32B	22.4 %	30A	7.6 %	49
North Carolina	39.8 %	12	1.0 %	38.8 %	15.7	14	19.7 %	9	25.2 %	10A	13.2 %	15
South Carolina	39.5 %	13	0.9 %	38.6 %	20.1	7	19.2 %	11	23.6 %	18B	112.0 %	20B
Iowa	38.9 %	14	5.9 %	33.1 %	12.1	22	11.3 %	41	23.5 %	20B	7.9 %	46A
Tennessee	36.4 %	15	1.6 %	34.8 %	23.1	4	21.6 %	4	24.9 %	13	13.4 %	13
Wyoming	34.8 %	16	3.1 %	31.7 %	6.9	46	10.0 %	46B	23.9 %	15	11.1 %	22B
Oklahoma	34.7 %	17	2.3 %	32.4 %	17.3	9	14.7 %	26B	25.2 %	10B	14.1 %	10
Alaska	34.3 %	18	0.2 %	34.1 %	20.5	6	9.5 %	49	27.2 %	4A	8.4 %	41

Table 1. Percentage of Rural Population, Fire Death Rates, and Other Explanatory Characteristics, by State (Continued)

State	Rural Population in 2000				Fire Deaths		Education		Smoking		Poverty	
	Percent Rural Population	Rural Population Rank	Farm Dwellers	Nonfarm Rural	per Million Population 1997-2001		Adults without 12 Years of School Average of 1998 and 2000		Adults Who Are Current Smokers 1999 percentage		People below Poverty Line, Average of 1998 to 2000	
	Percent	Rank	Percent	Percent	Average	Rank	Percent	Rank	Percent	Rank	Percent	Rank
Idaho	33.6 %	19	3.0 %	30.6 %	8.2	41B	15.6 %	19	21.5 %	37A	13.3 %	14
Wisconsin	31.7 %	20	2.6 %	29.1 %	9.4	30B	12.7 %	37	23.7 %	16B	9.0 %	39
Missouri	30.6 %	21	2.5 %	28.1 %	16.6	11	15.3 %	21	27.1 %	6A	9.8 %	34B
Nebraska	30.3 %	22	5.2 %	25.1 %	9.6	29	11.0 %	42A	23.3 %	22B	10.7 %	24
Indiana	29.2 %	23	2.1 %	27.2 %	14.4	17B	16.0 %	18	27.0 %	8	8.3 %	42
Minnesota	29.1 %	24	3.0 %	26.0 %	8.3	40	9.9 %	48	19.5 %	46	7.9 %	46B
Kansas	28.6 %	25	3.3 %	25.2 %	14.4	17A	11.4 %	40	21.1 %	40	10.5 %	26B
Georgia	28.3 %	26	0.8 %	27.6 %	15.8	13	18.7 %	14	23.7 %	16A	12.5 %	18A
Louisiana	27.3 %	27	0.7 %	26.7 %	21.3	5	20.3 %	8	23.6 %	18A	18.5 %	2
Virginia	27.0 %	28	0.9 %	26.1 %	12.5	19	15.4 %	20	21.2 %	39	8.1 %	43C
Michigan	25.3 %	29	0.9 %	24.4 %	14.9	16	14.2 %	29	25.1 %	12	10.2 %	29B
New Mexico	25.0 %	30	0.9 %	24.1 %	8.5	37	19.1 %	12	22.5 %	27B	19.3 %	1
Pennsylvania	23.0 %	31	0.7 %	22.3 %	15.1	15	15.1 %	23B	23.2 %	24	9.8 %	34C
Ohio	22.7 %	32	1.3 %	21.3 %	11.7	23	13.4 %	34	27.6 %	3	11.1 %	22A
Oregon	21.3 %	33	1.9 %	19.4 %	9.4	30C	13.2 %	35	21.5 %	37B	12.9 %	16
Delaware	20.0 %	34	0.6 %	19.4 %	11.6	24	14.4 %	28	25.4 %	9	9.9 %	33
Washington	18.0 %	35	0.8 %	17.2 %	9.4	30A	8.1 %	50	22.4 %	30D	9.5 %	37
Texas	17.5 %	36	0.9 %	16.6 %	11.4	25	21.3 %	5	22.4 %	30C	14.9 %	7
Colorado	15.5 %	37	1.1 %	14.4 %	5.3	48	10.4 %	45	22.5 %	27A	8.5 %	40

Table 1. Percentage of Rural Population, Fire Death Rates, and Other Explanatory Characteristics, by State (Continued)

State	Rural Population in 2000				Fire Deaths		Education		Smoking		Poverty	
	Rural Population	Percent Rural Population	Farm Dwellers	Nonfarm Rural	per Million Population 1997–2001		Adults without 12 Years of School Average of 1998 and 2000		Adults Who Are Current Smokers 1999 percentage		People below Poverty Line, Average of 1998 to 2000	
	Percent	Rank	Percent	Percent	Average	Rank	Percent	Rank	Percent	Rank	Percent	Rank
Maryland	13.9 %	38	0.5 %	13.5 %	10.9	26	14.8 %	25	20.3 %	43	7.3 %	50
New York	12.5 %	39	0.3 %	12.2 %	10.7	27	18.0 %	15	21.9 %	35	14.7 %	8B
Connecticut	12.3 %	40	0.1 %	12.2 %	8.1	43	14.1 %	30A	22.8 %	26	7.7 %	48
Illinois	12.2 %	41	1.0 %	11.1 %	12.2	21	15.2 %	22	24.2 %	14	10.5 %	26A
Arizona	11.8 %	42	0.2 %	11.7 %	8.4	38A	16.5 %	17	20.0 %	45	13.5 %	12
Utah	11.7 %	43	0.6 %	11.2 %	4.2	50	10.0 %	46A	13.9 %	50	8.1 %	43B
Florida	10.7 %	44	0.2 %	10.4 %	8.2	41A	17.1 %	16	20.7 %	41A	12.0 %	20A
Rhode Island	9.1 %	45	0.1 %	8.9 %	7.9	44A	19.0 %	13	22.4 %	30B	10.2 %	29C
Massachusetts	8.6 %	46	0.1 %	8.5 %	8.7	34A	14.7 %	26A	19.4 %	47	10.2 %	29A
Hawaii	8.4 %	47	0.5 %	8.0 %	5.1	49	14.0 %	32A	18.6 %	49	10.6 %	25
Nevada	8.4 %	48	0.2 %	8.2 %	8.7	34C	14.1 %	30B	31.5 %	1	10.1 %	32
New Jersey	5.7 %	49	0.1 %	5.5 %	8.7	34B	13.1 %	36	20.7 %	41B	8.1 %	43A
California	5.5 %	50	0.3 %	5.2 %	6.4	47	19.4 %	10	18.7 %	48	14.0 %	11

Note: Letters are used in the "Rank" columns to indicate ties and are assigned to States based on alphabetical order. For example, Alabama and New York both have poverty percentages of 14.7 percent, and so they are listed with ranks 8A and 8B, respectively.

Sources: NE-MW Economic Data from Northeast-Midwest Institute calculations based on data from U.S. Department of Commerce, Census Bureau, 2000 Census, Summary File 3, Table 3, Table P.5 Urban and Rural, data extracted via *http://factfinder.census.gov/*, accessed online at *http://www.nemw.org/poprural.htm* on May 23, 2005.

Hall, John R. Jr. U.S. Fire Death Rates by State. Quincy: National Fire Protection Association. October 2004, Table 3, pp. 9-12.

Incident types in rural and nonrural areas

From 1993 to 1995, the incident types for reported fires and civilian fire deaths were similar in rural areas and in the U.S. as a whole. Outside fires accounted for 45 percent of fires reported in rural areas and 43 percent of the reported fires in the entire Nation. Residential structure fires accounted for 25 percent of rural fires. Residential structure fires caused 69 percent of the rural civilian fire deaths and 60 percent of the rural civilian fire injuries. In the U.S. as a whole, residential structure fires accounted for 23 percent of the reported fires, 72 percent of the civilian fire deaths and 68 percent of the civilian fire injuries.

The graphs in this section reflect the data found in the USFA's report. [4]

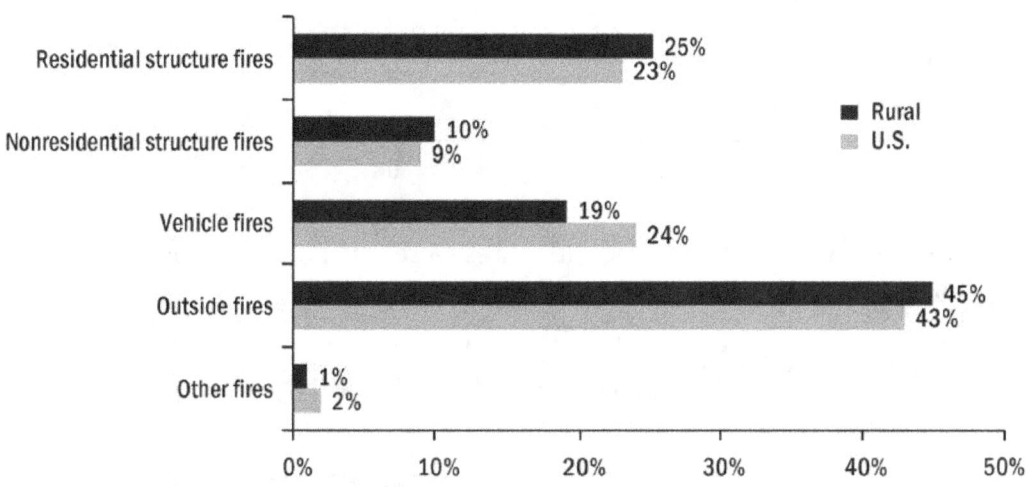

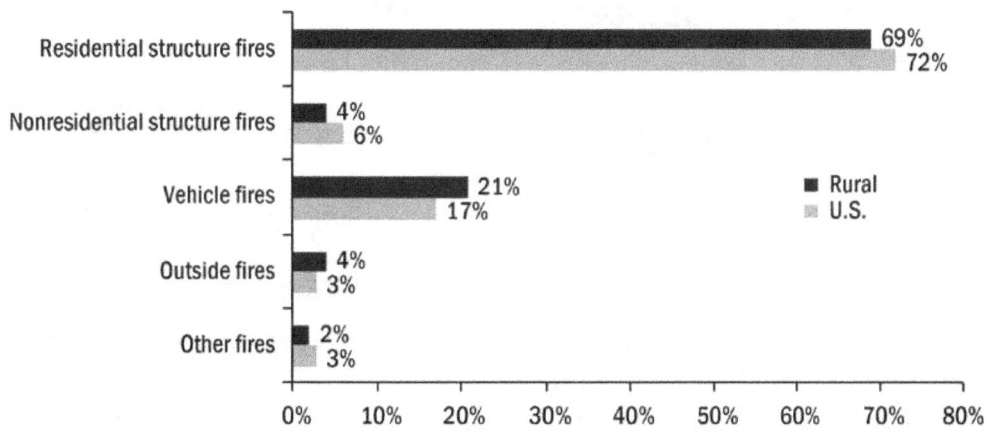

Cause profile is different in rural areas

Forty-five percent of the rural outside fires were caused by open flame, 16 percent by arson, and 9 percent by natural causes. In contrast, arson caused 44 percent of the nonrural outside fires.

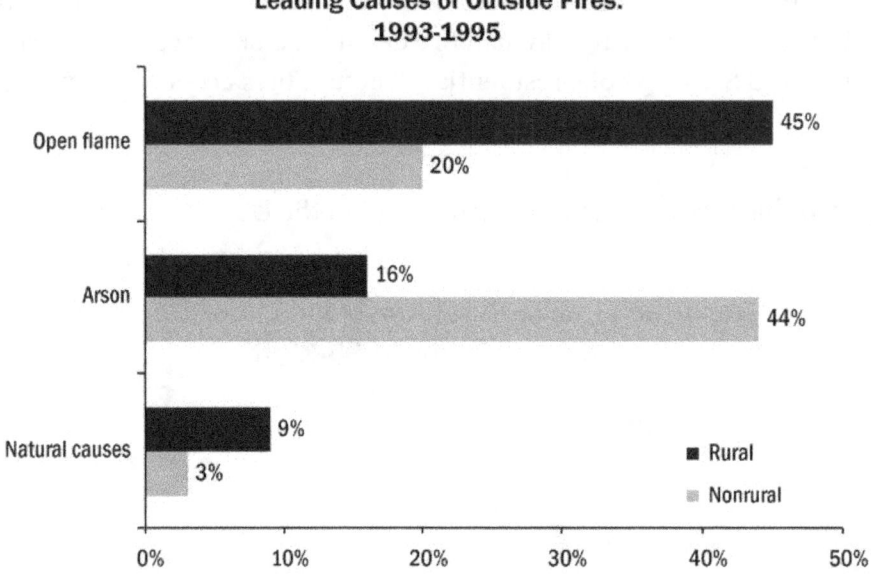

Thirty-six percent of the rural residential fires were caused by heating, 13 percent by cooking, and 12 percent by electrical distribution equipment.

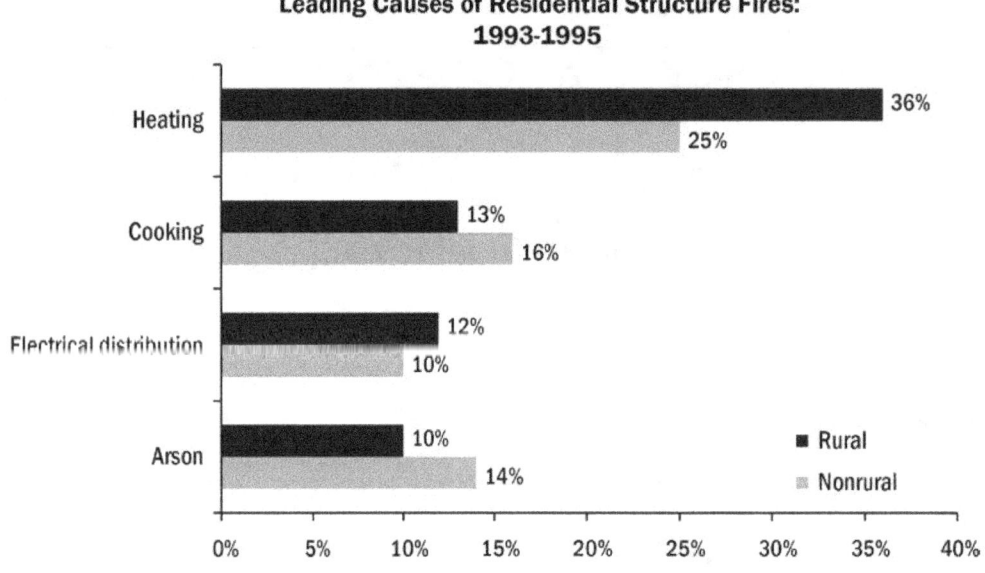

Twenty-six percent of the fatal residential rural fires were caused by heating, 23 percent by smoking, and 17 percent by electrical distribution equipment. Smoking caused 28 percent of the nonrural, fatal residential fires while 17 percent were caused by arson and 12 percent were caused by heating.

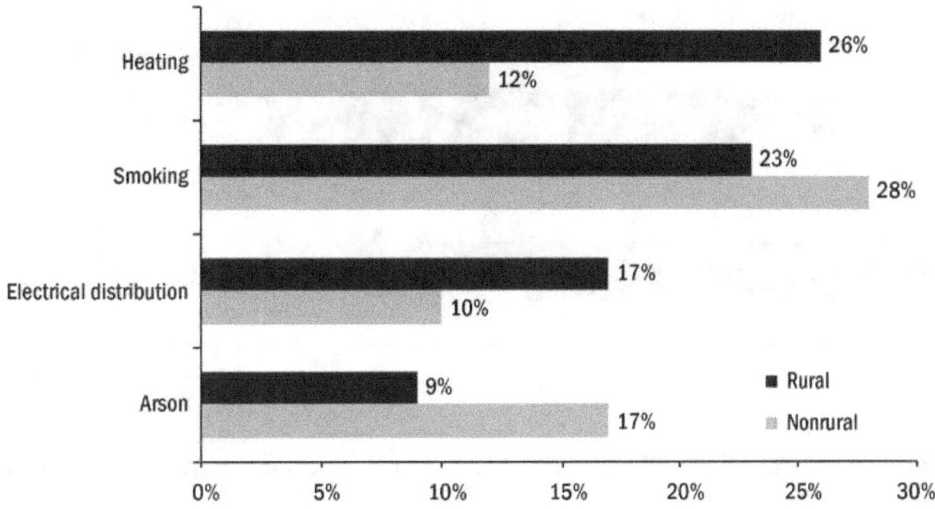

Fixed area heaters, including wood stoves, were involved in 38 percent of the rural residential heating fires. Chimneys (25 percent) ranked second, and fireplaces (11 percent) ranked third.

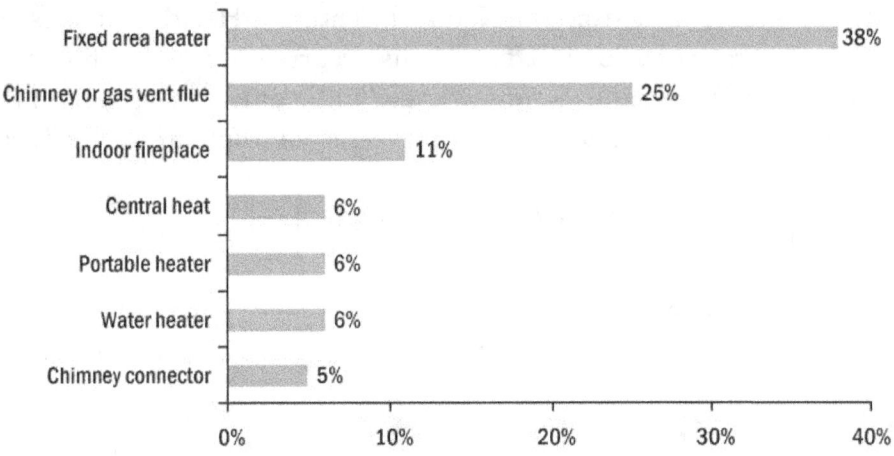

Smoke alarms

Almost three-quarters (73 percent) of rural residential fires occurred in properties without working smoke alarms compared to 65 percent in nonrural properties. The larger difference seen was in presence versus nonoperating. In 58 percent of the rural residential fires, no smoke alarms were present at all. In 15 percent of these fires, smoke alarms were present but not operating. In 42 percent of the nonrural incidents, no smoke

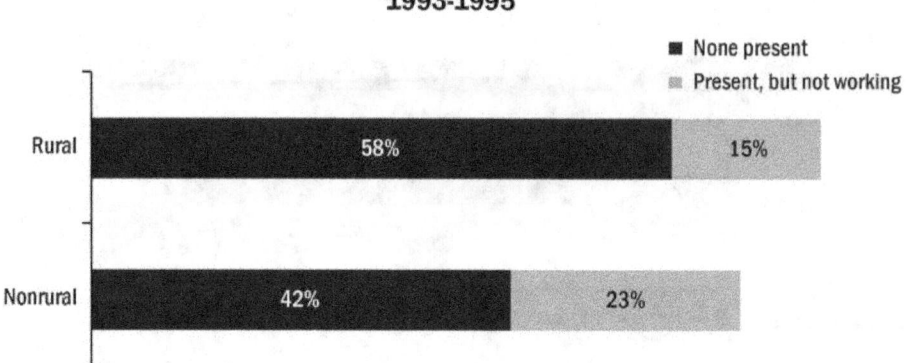

**Residential Fires with No Working Smoke Alarms:
1993-1995**

alarms were present at all. In 23 percent of these fires, smoke alarms were present but not operating.

Fire size in rural versus nonrural areas

Flame damage extended to the entire structure in 29 percent of the rural residential structure fires but only 17 percent of such incidents in nonrural areas.

Rural fire deaths by race

According to mortality data from the National Center for Health Statistics, during 1983-1988, an average of 5,764 U.S. fire deaths occurred per year. White victims accounted for 480 of the 676 rural victims per year. Little difference is seen in the percentage of fire victims by race or gender in rural versus nonrural areas. The racial picture

1983-1988 Fire Death Rates by Race

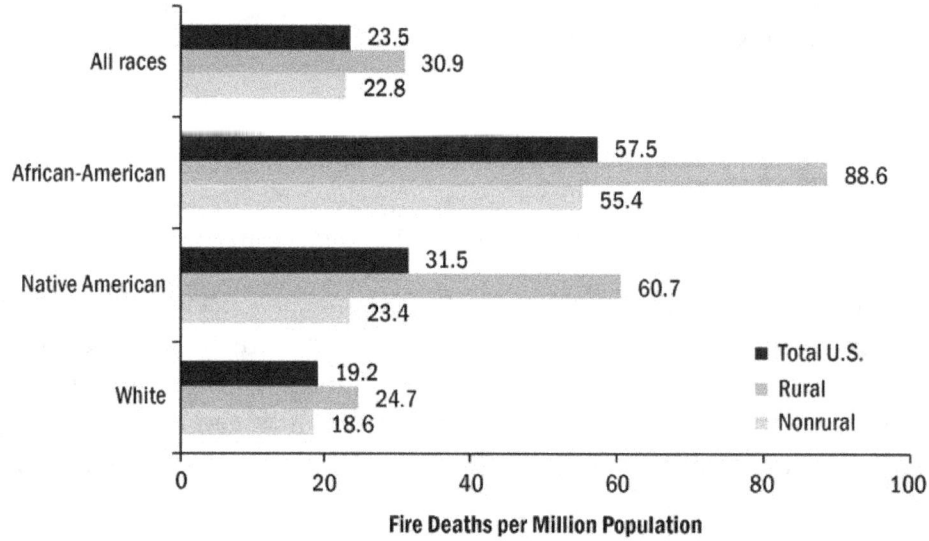

is different when death rates per million population are considered. From 1983 to 1988, the overall death rate for the U.S as a whole was 23.5 deaths per million. In rural areas, it was 30.9, and in nonrural it was 22.8.

Rural whites had a fire death rate of 24.7, rural Native Americans had a rate of 60.7, and rural African-Americans had a rate of 88.6, the highest of any group studied. The fire death rate for rural Native Americans was almost three times as high as the rate for non-rural Native Americans. Rural African-Americans had a fire death rate 60 percent higher than their nonrural counterparts. The fire death rate among rural whites was 33 percent higher than white nonrural residents.

References

1. Ahrens, Marty. *The U.S. Fire Problem Overview Report: Leading Causes and Other Patterns and Trends.* Quincy: National Fire Protection Association, Fire Analysis & Research Division, June 2003, pp. 27-31.

2. Northeast Midwest Institute. "2000 Rural Population as a percent of State Total By State," 2002, *http://www.nemw.org/poprural.htm*

3. Hall, Jr., John R. *U.S. Fire Death Rates by State.* Quincy: National Fire Protection Association, Fire Analysis & Research Division, Oct. 2004.

4. Federal Emergency Management Agency, U.S. Fire Administration. *The Rural Fire Problem in the United States,* Aug. 1997, *www.usfa.fema.gov/downloads/pdf/publications/rural.pdf*

Chapter 4. *Regional Differences*

Rural communities and the rural fire problem differ considerably from one U.S. region to another.

The South region (sometimes called the Southeast) is by far the most populous region (more than one-third of total U.S. population) and contains nearly half the total U.S. rural population. It is not uncommon therefore to talk about the rural fire problem and the South fire problem interchangeably. This is misleading. Rural communities have the highest fire incident and fire death rates in every one of the four regions in most years.

The West and the north-central regions contain some distinctive and important subgroups of the rural population that deserve separate attention but also should not be mistaken for the typical rural population of those regions. The West includes Native American communities, migrant worker communities, and Mexican border communities, sometimes called "colonias." The north-central region still has much of America's agricultural activity and farms.

The relative size and detailed characteristics of the rural fire problem differ depending on the region of the country. The U.S. Census Bureau divides the country into four primary regions: Northeast, north central, South, and West. (See Figure 1.)

Figure 1. Major Regions in the U.S.

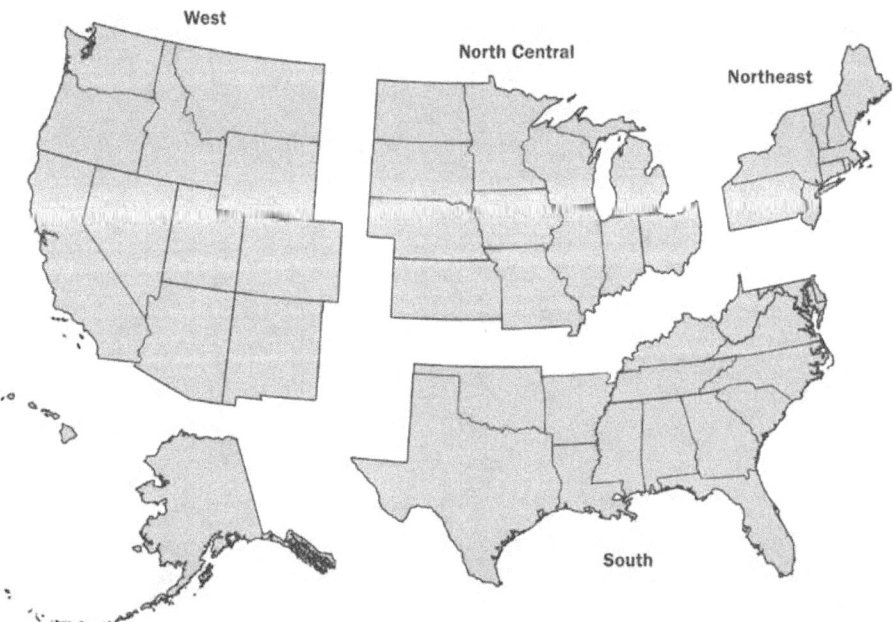

West

North Central

Northeast

South

The South has by far the largest share of the U.S. population (36 percent in 2000), and its rural share (26 percent in 2000) is larger than the rural share of any other region. The South accounted for 46 percent of the total rural population in the U.S. in 2000. [1]

In 3 of the past 10 years (1995 to 2004), the South's rural fire death rate was the highest of all the regions and the second highest in 5 of those years, making it the region with the largest number of years (8) in the top two. However, in the past 6 years, the South has had the highest regional rural fire death rate only once, which indicates that its rural fire death risk is now similar to that of other regions. From 2000 to 2004, the South had the highest rural fire death rate (29.0 deaths per million population) of all the regions by a narrow margin. [2]

The north-central region has the second largest share of the U.S. population (24 percent) and the second highest percentage of regional population in rural areas (28 percent). This means the north central has the second largest share of total rural population (28 percent). In the 1980s, rural areas were not consistently the community size with the highest fire death rates in the north-central region. [1]

In only 1 of the past 10 years (1995 to 2004), the north central's rural fire death rate was the highest of all the regions; it was the second highest in 4 years, making it the region with the lowest number of years as the highest region, but the second highest number of years (5) in the top two. In the past 6 years, the north central has had the highest regional rural fire death rate only once, and in 2000 to 2004 combined, the north central had the lowest rural fire death rate (22.8 deaths per million population) of all the regions. [2]

The Northeast has the fourth largest share of the U.S. population (20 percent) but the third highest percentage of regional population in rural areas (21 percent). This means the Northeast has the third largest share of total rural population (17 percent). [1] In the 1990s, rural areas were often not the community size with the highest fire death rates in the Northeast.

In 2 of the past 10 years (1995 to 2004), the Northeast's rural fire death rate was the highest of all the regions, and the second highest in 1 year, making it the region with the lowest number of years (3) in the top two. In the past 6 years, the Northeast has had the highest regional rural fire death rate only once, and in 2000-2004 combined, the Northeast had the second lowest rural fire death rate (27.0 deaths per million population) of all the regions. [2]

The West has the third largest share of the U.S. population (21 percent) but by far the lowest percentage of regional population in rural areas (14 percent), which is why the West has by far the lowest share of total rural population (12 percent). [1] However, these figures were as of 1990, and the West had by far the largest increase in nonmetropolitan population between 1990 and 2000 (21 percent compared to 12 percent in the South, 6 percent in the north central and 5 percent in the Northeast). [3] The West's share of total rural population has probably increased substantially, principally at the expense of the Northeast and north central regions.

In 4 of the past 10 years (1995 to 2004), the West's rural fire death rate was the highest of all the regions, but was never the second highest, making it the region with the highest number of years as top region but the second lowest number of years (4) in the top two. In the past 6 years, the West has had the highest regional rural fire death three times, and in 2000-2004 combined, the West had the second highest rural fire death rate (28.2 deaths per million population) of all the regions. [2]

Race and poverty

The rural South has nearly all of America's rural African-Americans and nearly all of America's poor rural African-Americans. In March 1997, 90 percent of nonmetropolitan African-Americans were located in the South, according to a 1998 Census Bureau study. [4] In 1996, 92 percent of nonmetropolitan African-Americans living below the poverty line were located in the South. [5]

African-Americans accounted for 14 percent of African-Americans and whites combined (excluding Hispanic whites) in metropolitan areas and 9 percent in non-metropolitan areas, in March 1997, nationwide. The corresponding combined national African-American percentage in March 1997, was 13 percent. The African-American percentage was 20 percent in metropolitan areas versus 19 percent in nonmetropolitan areas in the South, compared to 10 percent in metropolitan areas versus 2 percent in nonmetropolitan areas in all other regions combined.

In addition, in 1996, African-Americans accounted for 15 percent of African-Americans and whites combined (excluding Hispanic whites), and the national percentage of African-Americans living below the poverty line was 37 percent. In 1996 in the South, African-Americans constituted 52 percent of poor African-Americans and whites combined (excluding Hispanic whites) in metropolitan areas versus 42 percent in nonmetropolitan areas. In the same period using the same inclusion criteria (African-Americans and whites only and excluding Hispanic whites), the African-American percentages for the other three regions combined were 35 percent of the metropolitan poor and 4 percent of the nonmetropolitan poor.

And in 1996 for African-Americans and whites combined (excluding Hispanic whites), the percent of poor population was 8 percent for all metropolitan areas combined versus 15 percent for all nonmetropolitan areas combined; 12 percent for the metropolitan South versus 18 percent for the nonmetropolitan South; and 10 percent for the metropolitan part of all other regions combined versus 12 percent for the nonmetropolitan part of all other regions combined.

The implications of these statistics are that rural areas tend to be poorer than nonrural areas, but that difference is much more pronounced in the South. Rural populations tend to have a lower African-American share than do nonrural populations, but that difference is almost nonexistent in the South. The rural poor tend to have a lower African-American share than do the nonrural poor, but that difference is much more pronounced outside the South. In the rural South, when targeting the high-risk poverty population, white and African-

American each account for nearly half that target population. In rural areas outside the South, almost none of the high-risk poverty population is African-American (though because of the limitations of these statistics, there may be significant shares of Hispanic whites, Native Americans, or Asians in the poverty population outside the South).

The West has by far the highest Hispanic percentage in its population (24 percent in 2000 versus 12 percent in the South, 10 percent in the Northeast, and 5 percent in the north central). [6] The Hispanic population nationwide has the highest percentage of its people employed in farming, fishing and forestry (2.7 percent versus 0.5 percent for non-Hispanic white, 0.4 percent for African-American, 0.3 percent for Asian, and 1.3 percent for Native American). [7] These statistics seem to come as close as any readily available to substantiating and quantifying the importance of the Hispanic farming population, including migrant workers, in the rural West.

The West has by far the highest percentage of Native Americans in its population (2.0 percent versus 0.4 percent in the Northeast, 0.6 percent in the north central, and 0.8 percent in the South). [1] Native Americans are a high-risk group collectively and especially for those living on reservations. The West has by far the highest percentage of Native Americans on reservations in its population (0.7 percent versus 0.0 percent in the Northeast, 0.1 percent in the north central, and 0.4 percent in the South). Therefore, any rural programs for the West need to consider the distinct character of Native Americans, and reservations in particular, in their design and execution. Differences among different Nations (e.g., Navajo, Cherokee) also need to be considered as factors in customizing programs and delivery approaches. This applies to code enforcement programs as well. For example, the Navajo Nation does not have a fire code.

Related to the higher risk on reservations, Native Americans have the highest differential in fire death rate of any ethnic group when rural and nonrural areas are compared. For Native Americans, the fire death rate in rural areas is 2.6 times the rate in nonrural areas, compared to 1.6 for African-Americans and 1.3 for whites. [8]

Heating

The heightened share of rural fire deaths involving heating equipment is as much a South phenomenon as it is a rural phenomenon. For example, as noted, from 1993 to 1995, heating equipment accounted for 26 percent of residential structure fire deaths in rural areas versus 12 percent in nonrural areas, while smoking materials accounted for 23 percent of residential structure fire deaths in rural areas versus 28 percent in nonrural areas. From 1993 to 1997, heating equipment accounted for 14 percent of residential structure fire deaths in the Nation as a whole (rural and nonrural) compared to 23 percent for smoking materials. In the South (rural and nonrural) from 1993 to 1997, heating had a 19 percent share, the same as smoking materials. In the West (rural and nonrural), heating had a 10 percent share, compared to 22 percent for smoking materials. In the north central (rural and nonrural), heating had an 11 percent share, compared to 24 percent for smoking materials. And in the Northeast (rural and nonrural), heating had a 10 percent share, compared to 32 percent for smoking materials. [9]

Of the four regions, the South has the most consistently mild and short heating season. Therefore, poorer households in the South are the ones that find it most feasible to try to use space heating exclusively, resulting in the fire experience repeatedly documented for space heating as compared with central heating.

In 2003, use of central heating (warm-air furnace, steam, or hot water system) was highest in the north central region (90 percent) and Northeast (88 percent), where the entire region is subject to severe winters, while it was lower in the West (68 percent), where part of the region has consistently milder winters and part does not. Usage of central heating was lowest in the South (61 percent), where the entire region has consistently milder winters than the rest of the country.

Type of Primary Heating Equipment	Northeast	North Central	South	West
Warm-air furnace	40 %	81 %	59 %	65 %
Electric heat pump	1 %	2 %	24 %	6 %
Steam or hot water system	48 %	9 %	2 %	3 %
Floor, wall, or pipeless furnace	2 %	2 %	4 %	13 %
Built-in electric units	6 %	4 %	2 %	7 %
Room heaters with flue	1 %	1 %	2 %	1 %
Room heaters without flue	0 %	0 %	4 %	0 %
Stoves	1 %	1 %	1 %	2 %
Fireplaces	0 %	0 %	0 %	0 %
Cooking stoves	0 %	0 %	0 %	0 %
None	0 %	0 %	0 %	1 %
Portable electric heaters	0 %	0 %	1 %	1 %
Other	0 %	0 %	0 %	0 %
Total	**100 %**	**100 %**	**100 %**	**100 %**

Source: 2004-2005 *Statistical Abstract of the United States*, Table 947.

Some of these facts are puzzling. The West is closer to the South in heating equipment usage (central versus space), but the West's heating fire problem looks more like the Northeast and north-central regions, much lower than the South's. In addition, the leading type of space heating equipment in the South by far is electric heat pumps. Heat pumps, however, do not stand out as a specific type of equipment resulting in fires. These figures do not isolate rural dwellers or poor households, and it is possible the patterns are quite different for those high-risk groups.

Housing quality

Poor housing quality generally is a problem in the rural South. Three-quarters of the substandard housing units in the 1980s were in the South. In 1995, the 9 percent of the Nation's housing units in the nonmetropolitan South accounted for 21 percent of U.S. occupied units with moderate physical problems, 11 percent with severe problems, and 12 percent of the households with income below the poverty level. That last figure is consistent with the fact that the percentage of the population living in poverty, not limited to rural areas, is highest in the South (16 percent versus 13 percent in the West, 12 percent in the north central, and 11 percent in the Northeast, all in 1990). [10]

Manufactured homes

The South also has the highest proportion of housing units in manufactured homes—12 percent versus 3 to 7 percent in the other regions. [11] More than half (56 percent) of all the manufactured homes in the country are in the South. This has historically been a factor in the elevated fire death rate in the South, because until very recently manufactured homes have had a higher fire death rate than conventional "stick-built" homes or apartments. However, now that most manufactured homes in use were built after the advent of the construction requirements of the U.S. Department of Housing and Urban Development (HUD), introduced in 1976, manufactured homes are no longer a high-risk environment. This may be part of the reason why fire death rates in the South are no longer consistently much higher than rates in other regions.

The West has the second highest percentage of housing units that are manufactured homes (7 percent compared to 12 percent in the South, 5 percent in the north central, and 3 percent in the Northeast). As noted above, this may have been a factor increasing their rural fire death rate in the past but is probably not such a factor now.

Outdoor fires

The West has the highest percentage of fires and property damage due to fire involving outdoor properties (55 percent in 1995 to 1999 versus 51 percent for the South, 48 percent for the north central, and 47 percent for the Northeast). Outdoor fires in rural areas are likely to be unintentional involving open flame sources—a description that suggests open burning—while outdoor fires in nonrural areas are much more likely to be intentionally set.

Migrant workers and the "colonias"

Slesinger and Ofstead report that 159,000, or roughly 6 percent, of paid farm workers were migrants in 1985. Their study compares the characteristics identified by interviews with migrant Wisconsin workers conducted in the summers of 1978 and 1989. The workers tended to work in either the fields or the canneries. [12]

In 1989, 72 percent of the migrant workers in the area were male. Sixteen percent of the migrant men and 19 percent of the migrant women were functionally illiterate. Only 8 percent of the men and 14 percent of women migrant workers 25 years of age or older had completed high school compared with 76 percent of the U.S. population of that age. Sixty-two percent of the 1989 sample were married, and most married couples had children or other relatives as part of their household. Three generations were present in 11 percent of the households. Only 13 percent of the migrant workers described their heath as excellent, compared to 40 percent of the U.S. population as a whole.

While all of the migrant workers in the study lived away from home in July and August, roughly 90 percent were back in their home States in the winter months. Almost 60 percent were unemployed in the winter.

The median household income for these workers (average household of 5.2 persons) in 1988 was less than half the Federal poverty level and about a fifth of the national household median income. Migrant work was the sole income source for 44 percent percent of the households. During the peak season, workers often worked double shifts 7 days a week.

Most of the agricultural migrant worker lived in employer-provided housing. One-third of the housing units did not have indoor plumbing in 1989. In these cases, separate bath houses were provided. [12]

Susan Peck reports that the majority of farm workers in the U.S. are now Latino and that many of the workers are undocumented. [13] In the 1980s, farm workers in Delaware, Maryland, and Virginia were usually African-American or Caribbean, but as of 1992, 84 percent were Hispanic. African-American farm workers tend to be single men, while the Mexican and Mexican-American workers tend to travel as families. Many of these workers have no reading skills in any language.

In areas with long growing seasons, such as Oregon, farm workers often decide to stay, and the communities become more Hispanic. It is estimated that about 700,000 farm workers are hired annually in California, and 92 percent are foreign born. Just 9 percent were not authorized to work in the U.S. A growing number of farm workers are from the indigenous peoples in Mexico and Central America. Many of these people do not speak Spanish or English.

A 1993 report found that the median personal income for fieldworkers in California was between $5,000 and $7,500. Only 11 percent received food stamps, 2 percent were getting Aid to Families with Dependent Children, and 3 percent received housing assistance.

Increased enforcement of health, safety, and housing requirements and a corresponding increase in associated penalties have coincided with a decline in farmers providing housing for their workers. In 1968, there were 5,000 labor camps in California licensed by the State. In 1994, there were only 1,000. Nonemployer organizations are providing more and more housing. Labor contracting is becoming more common and direct hiring less so. [13]

Martinez, Kamaski, and Dabir describe the circumstances of the "colonias." [14] A "colonia," according to the definition used by the 1990 National Affordable Housing Authority Act is ". . . an identifiable community in Arizona, California, New Mexico or Texas within 150 miles of the U.S.-Mexico border, lacking decent water and sewage systems and decent housing, and in existence as a colonia before November 28, 1990." (p. 50) Other agencies and jurisdictions use different definitions. In 1995, the Texas Water Development Board estimated that roughly 280,000 people lived in 1,193 colonias in Texas, with 60 percent of this population in the four counties of the Lower Rio Grande Valley, counties that rank among the most impoverished in the country. Estimates of the number of New Mexico colonias and residents vary widely, ranging from 15 to 60 colonias, with 14,600 to 100,000 residents. In 1987, a Congressional Research Service (CSR) study found 25,000 in colonias of San Diego County and 11,500 in Imperial County colonias. Many rural Latino communities strongly resemble colonias but are outside of the border area. The 1987 CSR study found 50 to 55 colonias in Arizona.

Colonias do not fit neatly into traditional definitions of rural versus urban, are limited to four States, and are often ineligible for programs. Because colonias are physically, and generally legally, isolated, basic infrastructure such as water, sewer and paving lack economies of scale. Many colonias are not in cities and colonias' residents do not have sufficient income for user fees or many taxes. Colonia residents lack political power. They comprise only a small share of the local population in any voting district. Although advocacy by and services from community organizations were common in the colonias during the 1960s and 1970s, funding cuts had a major impact on these activities. [14]

References

1. U.S. Department of Commerce. *Statistical Abstract of the United States: 2006*, October 2005.

2. Karter, Jr., Michael J. *Fire Loss in the United States*. Quincy: National Fire Protection Association, Fire Analysis and Research Division, 1996-2005.

3. *Population Profile of the United States: 2000*, Figure 2-1, from *www.census.gov*

4. U.S. Census Bureau. Table 3, Distribution of the Population, by Region, Residence, Sex, and Race: March 1997, *http://www.census.gov/population/socdemo/race/black/tabs97/tab03.txt*, 1998.

5. U.S. Census Bureau. Table 15, Selected Characteristics of the Population Below the Poverty Level in 1996, by Region and Race, *http://www.census.gov/population/socdemo/race/black/tabs97/tab15.txt*, 1998.

6. Guzman, Betsy. *The Hispanic Population: Census 2000 Brief.* C2KBR/01-3, May 2001, *www.census.gov*

7. Fronczek, Peter, and Patricia Johnson. *Occupations 2000: Census 2000 Brief.* C2KBR-25, August 2003, at *www.census.gov*

8. Federal Emergency Management Agency, U.S. Fire Administration. *The Rural Fire Problem in the United States,* August 1997, *www.usfa.fema.gov/downloads/pdf/publications/rural.pdf*

9. Karter, Jr., Michael J. *U.S. Fire Experience by Region.* Quincy: National Fire Protection Association, Fire Analysis & Research Division, January 2001, Table 12.

10. U.S. Department of Commerce. *Statistical Abstract of the United States: 1994,* September 1994.

11. U.S. Department of Commerce. *Statistical Abstract of the United States: 2004-2005,* October 2004.

12. Slesinger, Doris P., and Cynthia Ofstead. "Economic and Health Needs of Wisconsin Migrant Farm Workers." *The Journal of Rural Health* 9, no. 2 (Spring 1993), pp. 138-148.

13. Peck, Susan. "Many Harvests of Shame: Housing for Farm Workers." *Housing in Rural America.* Joseph N. Belden & Robert J. Wiener, eds. California: Sage Publications, 1999, pp. 83-90.

14. Zixta Q. Martinez, Charles Kamasaki, and Surabhi Dabir, "The Border Colonias: A Framework for Change," *Housing in Rural America,* Joseph N. Belden & Robert J. Wiener, eds., California: Sage Publications, 1999, pp. 49-60.

Chapter 5. *The Rural Fire Service*

There are several principal distinguishing characteristics of the rural fire service. First, nearly all the fire departments are all or mostly volunteer. Second, because of low density of such communities, travel distances and travel times to fires and other emergencies tend to be longer. This means fire departments are more likely to have insufficient companies and personnel to meet national guidelines for effective response. In addition, fire departments are less likely to have needed equipment, and firefighters are less likely to have needed training; and fire departments are less likely to conduct fire prevention programs of all types, including code enforcement. Statistics in this chapter are taken from a 2001 needs assessment survey of the U.S. fire service, conducted by USFA and NFPA. [1]

Nearly all rural fire departments are all or mostly volunteer. In rural communities, there are a total of 13,440 fire departments, of which 43 (0.3 percent) are all-career, 32 (0.2 percent) are mostly career, 454 (3.4 percent) are mostly volunteer, and 12,911 (96.1 percent) are all volunteer.

Most fire departments have too few fire stations (and associated fire companies) to provide emergency response in compliance with the best available time and distance guidelines. Using maximum response distance guidelines from the Insurance Services Office (ISO) and simple models of response distance as a function of community area and number of fire stations, developed by the Rand Corporation, it is estimated that three-fourths to four-fifths of rural fire departments have too few fire stations to meet the guidelines.

In communities with less than 2,500 population, 21 percent of fire departments, nearly all of them all or mostly volunteer departments, deliver an average of four or fewer volunteer firefighters to a midday house fire. Because these departments average only one career firefighter per department, it is likely that most of these departments often fail to deliver the minimum of four firefighters needed to initiate an interior attack on such a fire safely.

Rural fire departments are less likely to have needed equipment, training, and other resources. Used vehicles accounted for an average of 42 percent of apparatus purchased by or donated to departments protecting communities with less than 2,500 population. Converted vehicles accounted for an average of 16 percent of apparatus used by departments protecting communities with less than 2,500 population. An estimated 15 percent of rural fire department engines (pumpers) are 15 to 19 years old, another 28 percent are 20 to 29 years old, and 22 percent are at least 30 years old. Therefore, two-thirds (65 percent) of all engines are at least 15 years old.

An estimated 34 percent of rural fire department fire stations are at least 40 years old, an estimated 72 percent have no backup power, and an estimated 92 percent are not equipped for exhaust emission control.

Rural fire departments do not have enough portable radios to equip more than half (49 percent) of the emergency responders on a shift. At least three-fifths of rural fire department portable radios (64 to 72 percent) are not water resistant, and at least two-thirds (69 to 84 percent) lack intrinsic safety in an explosive atmosphere.

An estimated one-half (48 percent) of rural fire department firefighters per shift are not equipped with self-contained breathing apparatus (SCBA). Half (53 percent) of their SCBA units are at least 10 years old. Two-fifths of emergency responders per shift (42 percent) in rural fire departments are not equipped with personal alert system (PASS) devices. An estimated 42,000 firefighters serving in rural fire departments lack personal protective clothing. Nearly half (45 percent) of their personal protective clothing is at least 10 years old.

An estimated 151,000 firefighters serving in communities with less than 2,500 population are involved in structural firefighting but lack formal training in those duties. An estimated 111,000 firefighters serving in communities with less than 2,500 population are involved in structural firefighting but lack certification in those duties.

An estimated 35 percent of rural fire department personnel involved in delivering emergency medical services (EMS) lack formal training in those duties. Two-thirds (65 percent) of rural fire departments do not have all their personnel involved in EMS certified to the level of Basic Life Support (BLS) and almost no departments (only 1 percent) have all those personnel certified to the level of Advanced Life Support (ALS). An estimated 50 percent of rural fire department personnel involved in hazardous material response lack formal training in those duties. Only 13 percent of rural fire departments have all their personnel involved in hazardous material response certified to the Operational Level and almost no departments (only 2 percent) have all those personnel certified to the Technician Level.

An estimated 45 percent of rural fire department personnel involved in wildland firefighting lack formal training in those duties. An estimated 56 percent of rural fire department personnel involved in technical rescue service lack formal training in those duties. An estimated 363,000 firefighters serve in rural fire departments with no program to maintain basic firefighter fitness and health.

Gaps in fire prevention programs

Rural communities are much more likely to have significant gaps in code enforcement than are larger communities. No one conducts fire code inspections in 39 percent of rural communities (less than 2,500 population), compared to 26 percent of communities of 2,500 to 4,999 population, 15 percent of communities of 5,000 to 9,999 population, 6 percent of communities of 10,000 to 24,999 population, and 0 to 1 percent of all larger communities.

Fire departments do not conduct plans review in 79 percent of rural communities, compared to 66 percent of communities of 2,500 to 4,999 population, 27 percent of communities of 5,000 to 9,999 population, 52 percent of communities of 10,000 to 24,999 population, and, at most, 15 percent of communities of all larger population sizes.

Permit approval, routine testing of active systems, and other code-enforcement activities show similarly large gaps.

Of the 61 percent of rural communities where someone does conduct fire-code inspections, the largest share for providers was "Other" with 22 percent, compared to 3 percent for full-time fire department inspectors, 11 percent for inservice firefighters, 12 percent for building inspectors, and 12 percent for a separate inspection department. The "other" category could involve any or all of such arrangements as State agencies, regional authorities, or contract inspectors.

Only these limited data are available to demonstrate the size of the gap for rural communities in the area of code-enforcement fire prevention and fire safety activities, and none of our interviews or literature reviews identified any examples of successful efforts by individual rural communities to address this gap in innovative ways.

With regard to fire prevention programs or activities other than those related to code enforcement, 80 percent of rural fire departments have no program for free distribution of smoke alarms, compared with 20 to 30 percent of departments protecting communities of at least 100,000 population. Rural communities and larger cities have the highest percentage of need for such programs, based on inference from their higher rates of poverty.

Juvenile firesetter programs are offered by 6 percent of rural fire departments compared to well over a majority of fire departments serving communities of 25,000 or more population.

Two-fifths (41 percent) of rural fire departments offer school fire safety education programs based on a national model curriculum, compared with 60 to 70 percent of fire departments serving all larger communities combined.

References

1. *A Needs Assessment of the U.S. Fire Service*. FA-240, FEMA, U.S. Fire Administration and National Fire Protection Association, December 2002.

Chapter 6. *Networks and Organizations in Rural America*

By building a fire safety program around existing networks, both the trust and the distance problems can be addressed. A network, as the term is used here, consists of a central organization with existing relationships for particular purposes with a larger group of people in the community. Several existing networks common to rural areas have been identified as potentially valuable to and supportive of fire safety programs. The central organizations for these identified existing networks are as follows:

- fire departments (nearly all volunteer in rural communities);

- health care (including both public health personnel and the individual private care providers, who may be the only ones located in a rural community);

- churches and other faith groups;

- schools;

- Fire Corps;

- area agencies on aging, senior citizen centers, Meals-on-Wheels and other older adult organizations (people age 65 and older are a high-risk group for fire death);

- rural electrical cooperatives;

- national safety organizations, such as the American Red Cross and Safe Kids Worldwide; and

- cooperative extension programs.

Farm-related groups were considered, but farm dwellers constitute only 1 percent of the U.S. population and only 5 percent of the rural population. Farm-related groups have only a fraction of the reach such groups had a century ago.

Although valuable to the success of fire safety programs, all of these networks have gaps in coverage. For example, fire departments are unlikely to have established relationships with their communities if their only activity is fire suppression. Some rural communities do not even have a fire department to call their own but rather obtain fire suppression services from a neighboring or regional authority.

A 2001 fire service needs assessment study (see Appendix B) found that 20 percent of rural fire departments had programs of free distribution of smoke alarms, compared to 40 percent of nonrural fire departments. Also, 41 percent of rural fire departments had school fire safety education programs, compared to 66 percent of nonrural fire departments. And only 11 percent of rural fire departments had in-service fire code inspections by firefighters, compared to 25 percent of nonrural fire departments. Other potential outreach programs showed similar large gaps between rural and nonrural departments.

These statistics suggest that rural fire departments will need to be recruited to serve as leaders or participants in expanded fire safety programs for their communities and that, if they agree, they will have to build up networks and relationships from a less advanced position than typically exists in larger communities.

The federally funded *Fire Corps* program, launched in December 2004, organizes community volunteers to help fire departments by performing nonoperational or non-emergency roles, such as fundraising, and public education activities, including education in local schools, home safety checks, and smoke alarm installation programs. Fire Corps is administered by the NVFC, with assistance from the IAFC. Because Fire Corps is part of a national network, there is support for the local Fire Corps to get involved in fire-safety education activities in local areas.

Area Agencies on Aging (AAAs) are in communities all across the country. They plan, coordinate, and offer services that help older adults remain in their homes aided by services such as Meals-on-Wheels, homemaker assistance, and whatever else it may take to make independent living a viable option. AAAs also work through local senior centers. Older adult organizations provide access to a portion of the 12 percent of the national population (or the somewhat higher percentage of the rural population) who are in the high-risk upper age groups.

While it is not known whether those who frequent senior centers include proportional participation by all ethnic, religious, education, and income groups, one must consider the possibility that participants will be drawn disproportionately from the lower-risk part of the age group. Fire service personnel will have to work with AAAs and other older adults organizations to make sure that not only are their programs distributed through senior centers, but also that they are reaching people in their own homes.

Schools provide access to the 5- to 17-year old age group (grades K to 12), which constitutes 18 percent of the U.S. residential population. However, this age group has below-average risk, except for the very youngest children. To be fully effective, school-based programs will probably need to provide access to other family members and/or provide significant effects for life. Neither of these larger effects has been well documented for even the most effective school programs.

Rural electrical cooperatives have a history that gives them potentially greater interest in the quality of life of their customers than one might expect from a typical power company, and that history also may give them a higher degree of existing trust. Presumably, those advantages of trust would be greatest for programs targeted on electrical equipment fires, because rural electrical cooperatives interact with their customers primarily on matters involving electrical equipment. Ordinarily, the targeting of electrical equipment fires would be hard to justify, because electrical distribution equipment ranks only sixth in share of fatal home fires, well behind smoking materials, intentional, heating equipment, and cooking equipment. However, in rural areas, the share of fatal fires involving electrical distribution equipment is higher than it is for nonrural areas. Therefore, this targeting **would** make sense within an overall strategy of targeting the larger parts of the rural fire problem.

Health care and faith networks probably have the broadest coverage of any of the existing networks. Like any network, these may vary from one to another in the degree to which their leaders and members see their missions broadly and strategically versus narrowly and traditionally. Individuals with a broad vision are more likely to be open to the value of participation in an ambitious fire safety program. Individuals with a narrow vision may still be willing to permit use of their resources in a more passive and less demanding fashion.

The U.S. Department of Agriculture (USDA) has networks through universities that have been granted federally owned land ("land-grant" universities). They have mandated outreach responsibilities, and they provide educational programs on a variety of subjects. A volunteer group that often works with them is the Extension Homemakers. The Extension Homemakers often get involved in safety education. There are extension offices in most counties throughout the United States.

Other national volunteer and service organizations such as the American Red Cross and Safe Kids Worldwide may have local or regional chapters that also reach small towns. Their outreach will vary throughout the country.

Because most of the existing community networks do not have a pre-existing interest in, or commitment to, fire safety program, the networks will have to be recruited to the cause.

The ideal situation is for a far-reaching and effective network to agree to full partnership and coleadership of a fire safety program. In other words, at least one network is needed that will agree to do the "heavy lifting" required for a program to be a success. At least one network must be found to agree to participate at this level.

Some networks may be unable or unwilling to commit to this level of participation but may be willing to recruit their members and constituents actively to the program. For example, a church might include in the weekly sermon a strong endorsement of the program. As another example, a health care facility might arrange for its doctors or other staff to encourage everyone they see in any capacity to participate in the fire safety program.

Finally, some networks may be willing to participate only in a passive manner. Imagine a church that allows third parties to place a program invitation on its bulletin board or in its newsletter.

Any help is welcome, of course, but a certain minimum level of help is necessary if a program is to be successful. A local network could be part of a national structure, and it may be that the local network can be approached effectively by way of its national counterparts. Many churches have a national hierarchy or organization, and health care facilities work with many national government agencies and nonprofit associations.

Finally, every community will have individuals and organizations that are not connected to a network but are potential sources for funds, other resources, and/or volunteers.

The perception of experts on fire and fire safety in rural America is that rural communities have a simpler, smaller group of key influential people. It is possible to ask who the most influential person is in town—the one best able to get things done—and receive a specific answer that proves accurate when tested. This means that when trying to set up a program by recruiting one or more networks and persuading those networks to take a substantial, demanding role, there may be only a few people who need to be sold on the proposition. In a smaller group, there is less bureaucracy and greater flexibility. The advantage is that there are fewer steps to success under these conditions. The risk, however, is that there may be no viable Plan B if the influential few are unconvinced.

In the long run, a smaller network can make a program more fragile, because the program's success will be so closely tied to the interest, energy, and continued involvement of a couple of people. Any disruption in that leadership core will be difficult to overcome. Therefore, it is essential to seek ways to broaden support and participation for a program early and often and to seek to institutionalize the program constantly in ways that will insulate it from variations in the fates and personal arcs of individuals.

Chapter 7. *Guidance for Effective Rural Safety Programs*

Fire safety professionals sometimes speak in terms of the three "E"s—education, engineering, and enforcement. Safety can be improved by changes in behavior, directly achieved by education, or indirectly achieved by well-enforced codes and standards, which cause people to change their behavior. Safety can be improved by engineered changes to living environments, compelled by well-enforced codes and standards, or through voluntary action. The following is guidance for volunteer firefighters, health and safety organizations serving rural communities, and other community leaders on how best to implement programs in rural areas.

1. **Pay particular attention to the program's ability to reach all or nearly all of its target audience. Rural communities have special challenges in program distribution.**

   In particular, rural communities have greater difficulty implementing fire safety changes by reaching target audiences through changes in codes and standards, door-to-door individual contacts, or mass media campaigns.

   A. **Programs that operate through local codes and regulations are challenged by gaps in code enforcement, which tend to be greater in rural communities.**

   Rural communities are much more likely to have significant gaps in code enforcement than larger communities. Fire code inspections are not conducted in 39 percent of rural communities (less than 2,500 population), compared to 26 percent of communities of 2,500 to 4,999 population, 15 percent of communities of 5,000 to 9,999 population, 6 percent of communities of 10,000 to 24,999 population, and 0 to 1 percent of all larger communities.

   Plans review, permit approval, routine testing of active systems, and other code-enforcement activities show similarly large gaps.

   B. **Programs that are delivered in person are challenged by the lower densities and greater place-to-place distances in rural communities.**

   The average distance between two points is four times higher in a nonmetropolitan area than in a metropolitan area. In addition, rural areas are the least dense areas in nonmetropolitan areas. Therefore, the distance ratio for rural versus nonrural may be even higher. In a door-to-door program, quadrupling distance means more time and more cost per household. Consequently, it may result in driving from place to place rather than walking from place to place.

C. **Programs that are delivered via mass media may be challenged by usage and access rates in rural communities for the selected mass media.**

Rural communities tend to have fewer television and radio stations within broadcast range. Satellite television may bring far more stations within range, but satellite television, like cable television, may be less affordable for the low-income rural residents who tend to be the primary targets for fire safety programs.

According to a 2003 survey by the U.S. Department of Commerce, rural households had closed the access gap with respect to the Internet. However, the gap was still large between rural poor and urban poor households.

Therefore, when planning for program delivery, it is important to consider the reach of selected mass media to target populations, which will tend to be the poorer members of the rural community. In addition, it is important to obtain up-to-date data on the reach of those media. Access has been increasing rapidly, and some perceptions of rural needs have been overtaken by events.

2. **Build a relationship of trust between those delivering a program and those targeted to receive the program.**

It is not clear whether the establishment of trust, an essential condition for program success, is more difficult in rural areas, as some believe. Some very successful fire safety programs in major cities have encountered trust issues in their target populations and solved them by working with and through faith networks.

Also, available data do not clearly support the perception of a higher degree of mobility and turnover in nonrural communities. The influx of strangers seems to be roughly as common in communities of all sizes.

Regardless of whether or not the establishment of trust is a major challenge elsewhere, it is viewed as a challenge in rural communities. A program without a trusted local advocate is unlikely to succeed.

Once the basis of trust has been established with some, however, there may be a very rapid spread of trust to the entire community if the perception is correct that there is a lower "tipping point" in a rural community.

3. **Build around existing networks.**

As previously discussed, several existing networks common to rural areas have been identified as potentially valuable to and supportive of fire safety programs. The ideal situation is for a far-reaching and effective network to agree to full partnership and co-leadership of a fire safety program. In other words, at least one network is needed that will agree to do the "heavy lifting" required for a program to be a success.

Other networks that may be unable or unwilling to commit to this level of participation may be willing to recruit their members and constituents actively to the program. For example, a church might include in the weekly sermon a strong endorsement of the program. As another example, a health care facility might arrange for its doctors or other staff to encourage everyone they see in any capacity to participate in the fire safety program.

Finally, some networks may be willing to participate only in a passive manner. For example, a church might allow third parties to place a program invitation on its bulletin board or in its newsletter.

It is also possible that a local network is part of a national structure, and it may be that the local network can be approached effectively by way of their national counterparts. Many churches have a national hierarchy or organization, and health care facilities work with many national government agencies and nonprofit associations.

Finally, every community will have individuals and organizations that are not connected to a network but are potential sources for funds, other resources, and/or volunteers.

4. **Try to work with a rural community's key influential people.**

    In order to get cooperation on getting things done it is important to work with people who have a history of getting things done in the community. These sometimes are heads of organizations, but many time people with an informal network—people who have a following. "Whenever meeting with someone ask the person you are meeting with "who do you go to when you want to get things done? Who else should I be talking with?" See Chapter 7 for more guidance.

5. **Plan for delays and setbacks, and be ready to adapt or respond as needed.**

    During the early stages of a program, it is especially important that program managers take steps to anticipate and plan for setbacks and other "bad news." The ability to respond quickly and effectively to any shock to the system is likely to be tested, and the need should be anticipated.

    A common theme in successful programs is the need for persistence, even doggedness. It takes persistence to enlist the influential few. It takes persistence to attract sponsors and partners. It takes persistence to gain access to the target population, even with pre-existing trust. Sustained effort, continuity, and followup are all phrases that came up repeatedly and with emphasis in our conversations with our experts.

6. Consult available books and other resources on best practices in the management of nonprofit organizations, with particular attention to the oversight and effective use of volunteers.

    The recruiting, training, and retention of sufficient numbers of workers, supervisors, and leaders are ongoing challenges to any program. Turnover is a major threat to program success.

    Societal trend that further complicate these challenges include declining participation in many kinds of volunteer activities, growing time pressures in modern life, and a tendency for rural dwellers to work somewhere outside their community. In addition, awkward and ineffective leadership succession arrangements—such as electing a new volunteer fire chief every year—can be problems.

    The use of rewards and recognitions, fundraising techniques, and management structures and procedures to address these challenges are all among the issues discussed at length in the literature on management of nonprofit organizations.

7. Modify model programs to reflect local conditions and provide local relevance.

    Information used to identify local conditions must be current and fact-based, and the changes should not damage the program elements that are essential to its design and effectiveness.

    Gathering the information can be done in a number of ways. The most assured but expensive approach is to ask the people in the community systematically. Less expensive versions of this approach include surveying a sample of the community, which can still be expensive, and interviewing a small group of opinion leaders, which can be misleading if those leaders do not know their constituents' opinions well.

    Regardless of how information is gathered, important subpopulations may be illiterate or speak only a language other than English, which will add to the complexity and cost of gathering their opinions. In addition, all the familiarity and similarity issues that affect trust and complicate program delivery also complicate any attempt to ask what people do and do not want.

    Compared to systematically asking people, passively listening is a less costly, more opportunity-driven approach that is also less intrusive and, as a result, less likely to create resistance among constituents. The downside is that this approach can be dominated by people with strong opinions, loudly expressed, and those people and their opinions may not be representative of the community.

    Whether opinions are obtained passively by listening or actively by asking, there is a value to the use of observation. The facts obtained through observation can be used to calibrate the accuracy of the inherently more subjective expressed opinions. In addition, a program whose messengers are also observers can build a database on a wide range of hazards, not just those that are the focus of the current program. This can provide direction to future expansion of the scope of the program, which can mean more safety and more impact for the community.

The approach least assured of success is reliance on pre-existing information regarding local preferences. The danger of this approach is that such information may be a collection of myths, stereotypes, and prejudices having only occasional correspondence with the community realities. These pre-existing beliefs also tend to be more general and more sweeping than the typically more complex reality, as the earlier-cited myth about Internet access illustrates.

Another complication in attempts to increase local relevance is that a community may not be the best judge of what it needs. Rural dwellers are not immune from stereotyping of their neighbors and, like people everywhere, they form their perceptions of what is most or least dangerous from a flow of information that is subject to a variety of distortions. Objective analysis of the correlation between fire experience and possible explanatory conditions or factors may show that the community's sense of what does and does not threatens it is not consistent with the facts.

This recommendation of customizing programs to local conditions puts a premium on the earlier-cited "influential few" and any other workers who have a detailed understanding of the constituents, what they like and do not like, what they need and want, and how to relate to them. The recommendation also places a premium on people who are good listeners and creative customizers but also thoroughly understand the programs and the reasons for their specific features and priorities. It also puts a premium on model programs that are designed for easy replication in different settings and easy but careful and sound customization.

Following are some common differences between the leading fire challenges in rural communities and those in other communities:

- Heating equipment accounts for a much larger share of fatal fires in rural communities than in nonrural communities. The share for electrical distribution equipment is also higher in rural communities, and the share for intentional fires (arson) typically is lower.

- Rural residents may be eligible for certain national government programs that provide grants or low-interest loans to rural residents (often limited to low-income households) to upgrade housing equipment for safety reasons. Some examples:

  — The U.S. Environmental Protection Agency has a Wood Stove Changeout Campaign (*http://www.epa.gov/woodstoves/changeout.html*) which provides financial incentives to replace older wood stoves with newer heating equipment, wood-burning or other, that pollutes less. Such a change could lead to reduced fire risk as well.

— The Rural Housing Service within the USDA has a program of grants and low-interest loans for low-income rural households for several related purposes. Upgrading of heating equipment is one of the purposes that could be eligible, and other repair projects that remove safety hazards appear eligible as well. The Rural Housing Service also has a program to assist rural fire departments and other emergency responder agencies to expand their ability to serve the needs of their communities.

— Rural electrical cooperatives have indicated a willingness to provide safety information to their customers and constituents not only on electrical system equipment and electric-powered appliances but on all types of household equipment.

- Within the outdoor fire problem, rural areas have their primary problem with outdoor burning, while urban areas have their primary problem with intentional fires.

- Water supplies can be a problem in rural areas. For firefighters, the absence of an onsite water supply suitable for firefighting, such as from hydrants connected to a piped public water supply under pressure, has several negative effects. One of these is to add a significant task of finding and accessing a local source for firefighting water. That task can require time and personnel who would otherwise be performing other tasks, it can delay the time when water is first put on the fire, and it can complicate the Command task for the officer in charge.

  — Sprinkler systems in homes also should be considered. Fire hoses, on average, use more than 8-1/2 times the water that residential sprinklers do to contain a fire. In remote areas, self-contained water tanks can be used to supply residential fire sprinkler systems. Therefore, when considering improvements to water supplies, consider options that will provide an infrastructure for more use of residential fire sprinklers as well as supporting firefighter hose operations.

8. To the maximum extent possible, once a fire cause or other topic has been selected as a focus for a rural fire safety program, use widely accepted safety messages that have been crafted and reviewed by experts for greatest effectiveness and greatest relevance to the specifics of that topic.

Here are the messages recommended by NFPA's fire and life safety experts for each of the major topics most likely to be selected as a focus for a rural fire safety campaign:

Educational Messages about Smoking

- If you smoke, use fire-safe cigarettes. These are cigarettes that go out if they are not actively smoked. They greatly reduce the risk of a cigarette fire. They are the only cigarettes permitted for sale in 20 States and Canada as of July 2007.

- If you smoke, smoke outdoors. There is less to catch fire outdoors, and if a fire starts, it is much less likely to endanger the people inside.

- Wherever you smoke, use deep, sturdy ashtrays. Ashtrays should be set on something sturdy and hard to ignite, like a table.

- Before you throw out butts and ashes, make sure they are out, and dowsing in water or sand is the best way to do that.

- Check under furniture cushions and in other places people smoke for cigarette butts that may have fallen out of sight.

Educational Messages about Heating

- Work with your fire department to find a professional who can review your heating equipment. Space heaters of all types need to be reviewed as well.

- When you buy a new space heater, select one that has the label of a recognized, independent testing laboratory. Make sure your community permits the use of the type of heater you choose. For example, portable kerosene heaters are illegal in some communities. And make sure the space you plan to heat is no bigger than the maximum space recommended for your heater by the manufacturer.

- Keep a 3-foot (1-meter) clearance between all heating equipment and anything that can burn. This means installing the equipment, if it is fixed, with clearance from items like the structural elements and the wall. It means placing portable heaters with clearance from fixed items and hard-to-move items, like a sofa or a bed. And it means always remembering to keep movable items, like clothing and papers, well away from the heating equipment.

- Turn off portable space heaters whenever you go to sleep or leave the room. Use central heating or additional bedclothes to stay warm at night. If you toss and turn at night you could toss bedding onto the heater, and if you get up in the middle of the night, you could knock the heater over.

- Plug electric heater power cords only into outlets with enough electrical power, not into extension cords. Power cords for electric heaters can draw a lot of current. Make sure you use an outlet with enough capacity for the heater and anything else you plug into that outlet. And never use an extension cord with a power cord.

- In your wood stoves, burn only seasoned wood that has dried for at least 12 months after it was split.   Do not use green wood, trash, or artificial logs. Start the fire with newspaper or kindling, never with any flammable liquid. Allow ashes to cool before disposal and place the ashes in a metal container kept a safe distance from the home and anything else that could burn.

- Make sure any fuel-burning heater—especially one that burns wood, coal, gas, or liquid—is vented to the outside in order to prevent a build-up of deadly carbon monoxide.

- In portable kerosene or other liquid-fueled space heater, always use the proper grade of the proper fuel. For example, do not burn gasoline in a heater designed only for kerosene.

Educational Messages about Electricity

- Have a licensed electrician check your home electrical wiring and distribution system, especially if your home is over 40 years old. If the electrician or you discover any electrical problems, shut off the circuit and have the problem fixed by a professional. Ask the electrician if arc fault circuit interrupters (AFCI) or ground fault circuit interrupters (GFCI) would help make your home safer.

What is an arc fault circuit interrupter?

— A new kind of circuit breaker, which shuts off electricity when a hazardous condition occurs.

— Older circuit breakers act when a circuit is overloaded with too much current, which can lead to overheating and start a fire.

— AFCIs act when they detect a low, irregular current, indicating that current is leaking out of the circuit.

— This leaking current is called an "arc fault" and it is an electrical discharge through damaged or missing insulation.

— If current leaks into things that can burn, a fire can start.

— If an arc fault occurs in wiring in the wall, the fire will be hidden and can spread a long way before smoke alarms can detect it. There may not be enough time for safe escape.

What can trigger arc faults?

— Damage to wires from nails driven into walls.

— Cracked wire insulation due to aging, stress, or rodents.

— Frayed or loose wires at stress points, such as where cords are flexed often or where plugs are removed by pulling on the wire and not the plug.

— Loose plugs or outlets.

— Faulty electrical appliances.

— Damage to insulation from overloading of circuits and resulting overheating.

> — Watch for signs of electrical problems, including:

> - An acrid odor near an outlet, switch, or light fixture,
> - Flickering lights,
> - Tripped circuit breakers or blown fuses, or
> - Outlet or switch that feels warm to the touch.

> Contact an electrician if you detect any of these signs of potential problems.

- Avoid overloading outlets and circuits.
- Limit the number and size of appliances plugged into a single outlet or a single circuit, to avoid overloading.
- Avoid use of extension cords as anything other than temporary.
- Use light bulbs with the right wattage. Check the maximum wattage rating on any fixture when you replace its light bulb.

Educational Messages about Outdoor Burning

- Make sure that outdoor burning of trash, brush, or other waste is done in accordance with local laws regarding permits, notification of authorities, and times and allowable ways of burning.
- Fire must be attended at all times.
- Avoid burning on windy, dry days. When conditions are windy or dry, it is too easy for open burning to spread out of control.
- Gasoline or other flammable or combustible liquids should not be used to burn trash, brush, or other waste.

Visible Address

- Make sure your house number can be seen easily from the street. Finding homes in rural areas can be difficult because of missing or blocked street numbers. To help firefighters or other emergency responders find your home, make sure there is a sign, a street number, or some other marker clearly visible form the nearest street.

9. **Consult the literature on rural safety and health program design and delivery.**

The project literature review identified two articles that had extensive guidance directly on point for any summary of best program practices. [1, 2]

The first article [1] reached these conclusions:

The term "rural" describes a variety of settings, including "frontier" counties with populations less than six per square mile and rural counties that abut urban ones. Differences also exist between farm and nonfarm rural populations.

Four core issues affect rural programs: geographic isolation, economic deprivation, human service infrastructure, and economies of scale. The distance that must be traveled is a part of the geographic isolation. Public transportation is close to non-existent. Older adults may feel that giving up driving is not an option due to a lack of alternatives. Terrain and weather also can make driving difficult. Costs for long-distance calls, fuel and travel time add up quickly.

Economic deprivation is exacerbated by the tendency for many rural areas to rely heavily on one industry, activity, or service for local livelihoods. Economic shifts or plant closures can be devastating to residents' income and local tax revenues. Also, some Federal and State programs mistakenly assume that services can be provided at less cost in rural areas, and do not fund adequately. Finally, because rural incomes are lower and fewer foundations are rural, there are fewer charitable resources available for programs or for the matching funds necessary to qualify for some grants.

The human service infrastructure has experienced consolidations and closings. There is a shortage of technical equipment and skilled personnel. Rural youth often move away and more women are working, reducing the volunteer pool that might partially alleviate the lack of paid workers.

Economies of scale are difficult because the numbers of people and suppliers are simply not there. In some cases, there is a sole supplier. Competitive bidding may not be possible.

In addition to these four core issues, rural housing is said to be less well maintained, rural residents less educated, and nonfarm rural residents tend to be in poorer health. Moreover, the independence associated with rural life, particularly among the elderly, often results in a resistance to using or accepting services or assistance. The following seven points should be considered when transferring urban programs to rural areas:

1) Expectations may need to be scaled back, particularly if success is defined as number of people served.

2) It is often necessary to scale services to offer only the highest priority (as defined by the community), rather than offering the full range.

3) Program duplication should be avoided and offerings coordinated so that each agency offers programs it can do best.

4) Rules and regulations should be handled with some flexibility as bookkeepers and accountants tend to be in short supply. Budget waivers should be sought when expenses will be higher than expected for items like long-distance calls and mileage.

5) Do not expect economies of scale or more than one provider bidding.

6) Create partnerships or reciprocal agreements so that the jurisdictional or administrative boundaries do not interfere with services.

7) Plan for challenges in recruiting and keeping qualified personnel. Hire people with multiple competencies rather narrow specialists.

Many services are delivered without benefit of formal office space. These services may be delivered from stores, churches, restaurants, or vehicles. Gatekeepers, including mail carriers, beauticians, and neighbors, can be used for referrals. The cooperative extension network is recommended as a vehicle for educational programs. Aging, nutrition, and hospital programs can support and publicize a new endeavor. Programs that operate in isolation are less likely to be successful.

The second article's [2] findings were as follows:

Distance, isolation, and resource shortages interfere with both problem identification and obtaining help with mental health issues. Ideally, an informal support system of family, neighbors, and friends helps rural elders through predictable life crises. Most services wait to be contacted, and traditional outreach tends to find those who are functioning fairly well. Several components common to successful direct service and educational programs are discussed.

In many mental health programs, one individual's enthusiasm, work, and commitment were critical in organizing and persuading others to establish a program. A credible "natural leader" from the community knows how to present the concept in a way that would be acceptable locally and could motivate other groups to participate.

In addition, services had to feel comfortable to clients. Impressions of comfort can be based on the program's appearance, location, time, and expense. Some programs provided services through nontraditional but trusted partners such as grain dealers, banks, and utility providers. Flexibility to choose to use portions of the services **when** they choose was important. Rural elderly women tended to be cautious about accepting formal services and were hesitant to accept when they felt that they could not reciprocate.

Two direct service models, gatekeepers and peer counseling, were described in this second article. Gatekeepers routinely have contact with people who themselves would not seek services. These gatekeepers make referrals to appropriate agencies. Peer counseling uses trained, elderly volunteers from the same community as part of the mental health team. Elders were more open to their own peers than they were to professionals whom elders fear might threaten their independence. Peers often have a more realistic understanding of the client's situation.

Three educational program models that trained nonprofessionals to recognize mental health problems and make appropriate referrals also were discussed. The nonprofessionals received cross-training about a number of organizations. The organizations had to gain a better understanding of the mental health needs of rural older adults and to collaborate in service provision. The three programs discussed were as follows:

- A Missouri project trained trainers, using the well-established networks of area agencies on aging and university extensions. At eight sites in the State, local committees comprised of representatives of these two organizations, and community leaders, including clergy, community mental health providers, elder volunteers, or senior center directors, coordinated recruiting of candidate trainers who wanted training in mental health and aging. The project a) improved the providers' ability to recognize mental health needs, b) identified rural older adults who needed service, c) provided information on referral programs, d) assisted providers in notifying possible clients about their services, and e) helped providers inform agencies about possible clients.

- An Arizona program presented workshops for rural health (not mental health) staff, volunteers, family, and community members on identifying signs of mental illness in older adults, how to respond more effectively to people with these problems, and appropriate referrals. Culturally specific curricula were developed for rural Anglos, Latinos, and Native Americans. The local supports helped continue training after the program ended.

- A Pennsylvania program used a cross-system training model and focus groups to create a curriculum. Volunteers, gatekeepers, and nonprofessional caregivers were trained to give educational presentations in their communities. Recipients of training then conducted more training. This approach brought staff of the mental health and aging systems together at meetings.

All three programs used committees and focus groups with community leaders in formal and informal roles. The committees helped develop the curricula specific to their locations and played key roles in marketing the program and recruiting people to participate. In addition, the importance of rural-specific material in such programs was stressed. Crisis intervention materials that advise calling 9-1-1 are not appropriate in areas that lack that service. Instead, the materials need to provide contact information for the appropriate local emergency contact, who may be the sheriff.

References

1. Bull, C. Neil, and Shari DeCroix Bane. "Program Development and Innovation." *Journal of Applied Gerontology*, vol. 20, no. 2, June 2001, pp. 184-194.

2. Bane, Shari DeCroix, and C. Neil Bull. "Innovative Rural Mental Health Service Delivery for Rural Elders." *Journal of Applied Gerontology*, vol. 20, no. 2, June 2001, pp. 230-240.

Chapter 8. *Recommendations*

The following are recommendations for further research and for implementation of programs for USFA, NFPA and other national, State, and local organizations interested in mitigating the rural fire problem.

1. The project found that the rural fire problem involves larger shares for heating equipment and electrical distribution and lighting equipment fires than in other communities. In most cases, reductions of these risks will involve some updating or upgrading of heating or electrical distribution equipment. These improvements go beyond what is required in current codes and standards, and will be difficult for the highest-risk rural households to afford on their own.

    We recommend the development and implementation of a model multihazard survey for homes that could be incorporated as a voluntary outreach program and used to identify homes that need changes in their equipment. This survey would be similar to a battery of tests when you go to your health care provider for an annual physical. Equipment checked could include both portable and stationary space heaters, electrical wiring and related parts of the electrical distribution system, and smoke alarms. The survey also could check related conditions, such as locked, blocked, or inoperable doors and windows that are part of primary or alternate escape routes.

    For greatest effectiveness and least burden on the households, the survey would be conducted by trained professionals, though not necessarily certified fire inspectors or electricians, with the consent of the households. Despite the term "survey," this is not envisioned as a hand-off instrument for households to use to review their own equipment.

    After the survey, the residents of the household would be given a list of prioritized safety hazards that should be corrected. In an ideal program, there would be community block grant or other funding that would help the property owner to follow through on some of the improvements suggested by the survey.

2. As an enhancement to recommendation #1, we further recommend the production of a walk-through video showing a home survey in a rural home. This video will help volunteers know where to look for equipment, hazards, or other conditions on the list of items to be reviewed, and it will help volunteers recognize different types of equipment and the visible signs of hazardous or unsafe conditions. This video can be used as a training vehicle with local volunteers who plan on running a survey, can be shown on local cable access television for rural home owners, or can be shown at meetings in local venues such as senior centers or in rural electric community halls.

3. We recommend partnering with national and regional organizations and agencies, such as the U.S. Environment Protection Agency (EPA), the Southwest Indian Foundation; the USDA; the Hearth, Patio and Barbecue Association; and the HEARTH Education Foundation, to develop programs that would replace problem space heaters. Such a program is recommended if the hazardous equipment or conditions identified in the multihazard surveys are to be addressed.

    Many of these organizations and agencies already provide some kind of heating improvement program. The EPA has a wood stove change-out campaign that provides financial incentives to replace older wood stoves with newer heating equipment, including wood stoves that pollute less. (See *http://epa.gov/woodstoves/ how-to-guide.html* for more details.) It should be straightforward to modify this program to swap in wood stoves or other heating equipment that also poses less fire risk. The USDA's Rural Housing Service also has program grants and low-interest loans that include upgrading of heating equipment.

4. We recommend development of a program for improvement of rural electrical system safety that will set priorities in terms of the range of hazards and conditions that may be identified in a survey and will identify affordable modifications suitable for use in existing homes. Priorities may involve direct indicators of problems, such as an acrid odor near an outlet, switch, or light fixture; flickering lights; or tripped circuit breakers or blown fuses. Priorities also may involve general conditions known to be correlated with higher risk, such as an older electrical system or the use of certain wiring materials (e.g., aluminum) or methods.

    We recommend partnering with the National Rural Electric Association (NREC) the Electrical Safety Foundation Institute (ESFI), the Consumer Product Safety Commission (CPSC), Underwriters Laboratories (UL), and the NFPA for both technical expertise and assistance with implementation. Such a program is recommended as a complement to the multihazard survey in order to create a politically and economically realistic strategy for addressing the electrical hazards identified in the surveys.

5. Recent surveys indicate that smoke alarms are now in nearly all American homes (about 96 percent in the most recent surveys) and that the percentage usage in rural America as a whole is not far behind (at least 90 percent). In the highest-risk parts of rural America—such as the Southeast and the First Nation communities—there are often large gaps. Smoke alarms are probably the best combination of affordability and life safety effectiveness available, and any program to improve fire safety in rural America needs to include a strong component to achieve universal smoke alarm usage. Furthermore, research has shown that the most effective programs are those that not only donate smoke alarms but also install them. This suggests incorporation of smoke alarms as the centerpiece of a multihazard survey program such as that outlined in recommendation #1, but because not every rural community will be in a

position to immediately pursue recommendation #1 in its entirety, there needs to be a strong model program for places where only the smoke alarm gap can be addressed immediately.

We recommend the development of a national strategy to install working smoke alarms in every rural home. This strategy can begin by working with national organizations that already have proven programs, such as CDC's National Center for Injury Prevention and Control and the NFPA to advocate for more funding, more coordination, and planning sufficient to reach the goal of universal coverage in a reasonable period of time. In 2003 and 2004, NFPA/USFA conducted an installation program reaching nearly 4,000 residences in Holmes County, Mississippi, one of the poorest counties in the U.S. Surveys of installers working in Holmes County reported that only 53 percent of the homes had at least one smoke alarm before the installers put in new alarms. But if the total U.S. gap in smoke alarm coverage is about 4 million households and the rural gap is therefore just under 1 million households, dozens of programs like the Holmes County project will be needed if this ambitious goal is to be achieved in the next decade.

6. The elevated fire risks in America's rural areas remain largely unrecognized among much of the population, and the great diversity among rural communities is an obstacle to perception of this great problem. America needs to understand the rural fire problem better, but rural America needs to see their shared exposure to risk hidden beneath their diversity of living circumstances.

    Toward that end, we recommend the development of a DVD/video that would communicate the importance of reaching rural communities and would portray the variety of rural communities in the United States by region and group type. This DVD footage would show the sparseness of rural communities and visually set the rural communities apart from other larger communities. It would feature interviews with people from rural areas in the Northeast, the Midwest, the West, the Southeast, migrant communities, and First Nation communities. This video could be used at national, regional, and local meetings to motivate people to get involved with mitigating the rural fire problem and to motivate rural dwellers to work with each other.

7. A truly national campaign to reduce the rural fire problem is unlikely to develop solely through the independent, self-starting efforts of the 13,750 communities of under 2,500 population. Even if there are not enough resources to achieve a massive infusion of people and materials to every community, there is a national role in providing model programs, guidance in building program networks, and help in coordinating with other communities.

    We recommend development of organizational options for providing a supportive network that could be extended to every rural community. In particular, consider how multiple levels of program oversight might be defined. Suppose there was a network of first-level coordinators for the 13,750 community program managers.

Based on estimates of the number of hours a month each community will need from a first-level coordinator and how many hours a month each coordinator will need for other duties (e.g., development of materials), one can calculate how many first-level coordinators will be needed to cover the country. If the number is large enough, one can estimate how many second-level coordinators are needed for the first-level coordinators, and so forth. It would be worth some research to determine whether any similar national effort has been mounted and carried forward successfully.

8. Fire service needs assessments have shown extensive needs for all types of resources, and invariably a larger percentage of rural fire departments have needs than do fire departments in any other size community. This includes the need for sufficient fire stations and firefighters to provide emergency response with enough speed to be effective and enough people to be safe while operating.

    Moreover, the number of volunteer firefighters has been down, not up, for most of the past two decades. Even if some of this decline represents conversion of fire departments from volunteer to career (the trend has been generally up for career firefighters), it seems clear that the significant needs for more firefighters in rural fire departments are not being met in the natural unfolding of events.

    We recommend increased research on effective ways to meet the needs of the rural fire service. New ways to recruit and retain volunteer firefighters would be very useful in this effort.

Appendix A. *Literature Review and Statistical Analysis*

Defining "rural"

In any discussion of the rural fire problem, the first question that must be asked, is "What do we mean by rural?" This sounds deceptively simple.

The Rural Assistance Center notes that the U.S. Census Bureau, the Office of Management and Budget, and the Economic Research Service of the USDA all have definitions that are widely used. The U.S. Census Bureau considers areas other than urbanized areas (population of 50,000 or more) or urban clusters (population of 2,500 to 49,999) to be rural. The Census Bureau definition of rural, therefore, includes all and only, those communities with populations of less than 2,500. This is the primary definition used in this analysis.

However, many of the studies referenced use other definitions. The Office of Management and Budget separates counties into a) metropolitan area that include counties with urbanized areas and economically connected outlying counties, and b) nonmetropolitan areas. Nonmetropolitan areas often include micropolitan areas with at least one urban cluster of 10,000 or more residents. The Economic Research Service (USDA) defined Rural-Urban Commuting Areas (RUCAs) using census tract classifications, census definitions of urbanized area and place, and commuting information. [1,2] The American Heritage dictionary defines rural as "of or relating to the country rather than the city."

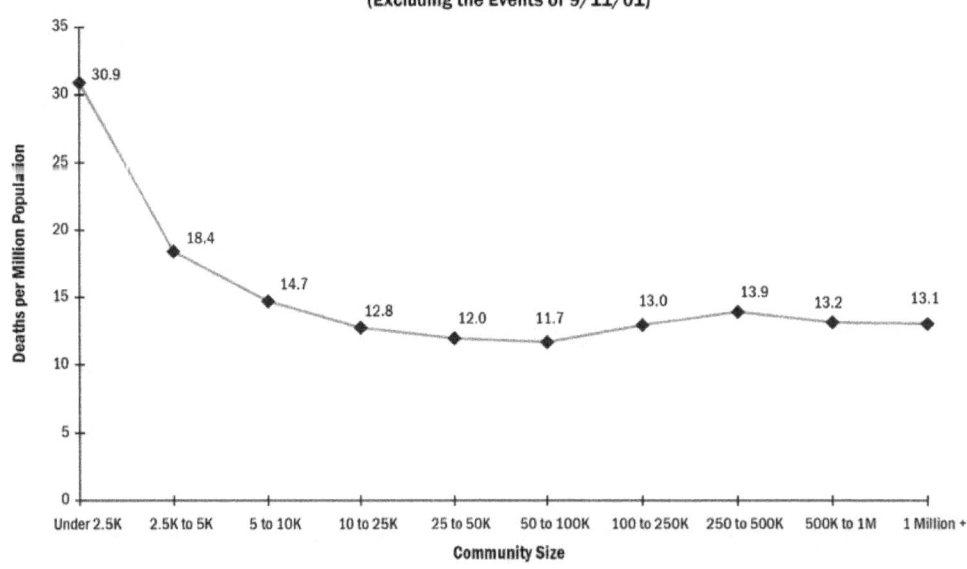

Civilian Fire Deaths per Million Population by Size of Community: 1997-2001 Annual Averages
(Excluding the Events of 9/11/01)

Source: *Fire Loss in the United States,* by Michael J. Karter, Jr.

The size of the rural fire problem

During the five-year period 1997 to 2001 (excluding the events of September 11, 2001), rural communities with populations under 2,500 had an average fire death rate of 30.9 per million population. This rate was at least twice that found in most other population intervals with the exception of communities with populations of 2,500 to 5,000, which had a rate of 18.4 fire deaths per million population. Communities with populations under 2,500 averaged 12.0 reported fires per 1,000 population, twice that of all population intervals except 2,500 to 5,000 (8.3), and 5,000 to 10,000 (6.9). These communities also have the highest per capita rate of reported fires. [3]

According to U.S. Census figures assembled by the Northeast Midwest Institute, 59 million, or 21.0 percent, of the U.S. population in 2000 lived in rural areas. Three million, or 1.1 percent, of the U.S. population lived on farms. Fifty-six million, or 19.9 percent, lived in nonfarm rural areas. [4]

Table 1 shows the percentage of rural, farm, and nonfarm rural populations and rank by rural percentage for each State in 2000. These data are combined with data from a table in John Hall's 2004 report, *U.S. Fire Death Rates by State,* showing State fire death rates and rank for the 5-year period 1997 to 2001, along with explanatory characteristics and their ranks. Vermont and Maine ranked first and second of percent of rural populations, but their fire death rates were 12th and 20th, respectively. Among the 15 States with the largest percentages of rural populations, 9 were among the 15 States with the highest fire death rates.

The percentage of rural population is a powerful statistical predictor of fire death rates. Hall used simple linear regression analyses to estimate the strength (in percent of variation explained) of candidate variables. Lack of education, defined as the percent lacking 12 years of school, accounted for 29 percent of the variation and was the strongest of the three predictors discussed. The percent of current smokers accounted for 22 percent of the variation and the percent below the poverty line accounted for 16 percent. Using the same techniques, percent rural explained 31 percent of the variation.

Most high fire death rate States are in the South and have large percentages of rural populations. However, Alaska is the northernmost part of the U.S., and its fire death rate has consistently been high.

Hall found that almost all States have shown drops in fire deaths and fire death rates over the past two decades. The Southeastern States of the old Confederacy (excluding Florida), plus Alaska, have consistently had fire death rates above the national average. Border States such as Missouri and Oklahoma also tend to have high rates. He noted that States with small populations can have unusually high death rates in some years due to an increase of just a few deaths. [5]

The rural population in the U.S is not homogenous. Before attempting to address the rural fire problem, it is necessary to understand something about the people, lives, and regional differences of these rural populations. Fire, injury, and sociological literature were searched for rural references. This literature review includes descriptions of rural populations in general, as well as specific rural populations. Fire and injury patterns are discussed. In addition, some public health programs also are reviewed as possible models.

Table 1. Percentage of Rural Population, Fire Death Rates, and Other Explanatory Characteristics, by State

State	Rural Population in 2000				Fire Deaths		Education		Smoking		Poverty	
	Rural Population Percent	Percent Rural Population Rank	Farm Dwellers Percent	Nonfarm Rural Percent	per Million Population 1997-2001 Average	Rank	Adults without 12 Years of School Average of 1998 and 2000 Percent	Rank	Adults Who Are Current Smokers 1999 Percentage Percent	Rank	People below Poverty Line, Average of 1998 to 2000 Percent	Rank
Vermont	61.8 %	1	1.8 %	60.0 %	16.2	12	11.7 %	39	21.8 %	36	10.3 %	28
Maine	59.8 %	2	0.9 %	58.9 %	12.3	20	12.0 %	38	23.3 %	22A	9.8 %	34A
West Virginia	53.9 %	3	1.2 %	52.7 %	18.1	8	23.3 %	1	27.1 %	6B	15.8 %	4B
Mississippi	51.2 %	4	1.6 %	49.6 %	32.1	1	21.2 %	6	23.0 %	25	15.5 %	6
South Dakota	48.1 %	5	7.7 %	40.4 %	9.1	33	11.0 %	42B	22.5 %	27C	9.4 %	38
Arkansas	47.6 %	6	1.9 %	45.6 %	24.2	3	20.8 %	7	27.2 %	4B	15.8 %	4A
Montana	46.0 %	7	4.4 %	41.5 %	10.2	28	10.7 %	44	20.2 %	44	16.0 %	3
Alabama	44.6 %	8	1.2 %	43.3 %	25.9	2	21.9 %	2	23.5 %	20A	14.7 %	8A
Kentucky	44.3 %	9	3.2 %	41.1 %	17.1	10	21.7 %	3	29.7 %	2	12.5 %	18B
North Dakota	44.2 %	10	6.8 %	37.4 %	8.4	38B	15.1 %	23A	22.2 %	34	12.8 %	17
New Hampshire	40.8 %	11	0.4 %	40.4 %	7.9	44B	14.0 %	32B	22.4 %	30A	7.6 %	49
North Carolina	39.8 %	12	1.0 %	38.8 %	15.7	14	19.7 %	9	25.2 %	10A	13.2 %	15
South Carolina	39.5 %	13	0.9 %	38.6 %	20.1	7	19.2 %	11	23.6 %	18B	112.0 %	20B
Iowa	38.9 %	14	5.9 %	33.1 %	12.1	22	11.3 %	41	23.5 %	20B	7.9 %	46A
Tennessee	36.4 %	15	1.6 %	34.8 %	23.1	4	21.6 %	4	24.9 %	13	13.4 %	13
Wyoming	34.8 %	16	3.1 %	31.7 %	6.9	46	10.0 %	46B	23.9 %	15	11.1 %	22B
Oklahoma	34.7 %	17	2.3 %	32.4 %	17.3	9	14.7 %	26B	25.2 %	10B	14.1 %	10
Alaska	34.3 %	18	0.2 %	34.1 %	20.5	6	9.5 %	49	27.2 %	4A	8.4 %	41

Table 1. Percentage of Rural Population, Fire Death Rates, and Other Explanatory Characteristics, by State (Continued)

State	Rural Population in 2000				Fire Deaths		Education		Smoking		Poverty	
	Percent Rural Population Percent	Rural Population Rank	Farm Dwellers Percent	Nonfarm Rural Percent	per Million Population 1997-2001 Average	Rank	Adults without 12 Years of School Average of 1998 and 2000 Percent	Rank	Adults Who Are Current Smokers 1999 Percentage Percent	Rank	People below Poverty Line, Average of 1998 to 2000 Percent	Rank
Idaho	33.6 %	19	3.0 %	30.6 %	8.2	41B	15.6 %	19	21.5 %	37A	13.3 %	14
Wisconsin	31.7 %	20	2.6 %	29.1 %	9.4	30B	12.7 %	37	23.7 %	16B	9.0 %	39
Missouri	30.6 %	21	2.5 %	28.1 %	16.6	11	15.3 %	21	27.1 %	6A	9.8 %	34B
Nebraska	30.3 %	22	5.2 %	25.1 %	9.6	29	11.0 %	42A	23.3 %	22B	10.7 %	24
Indiana	29.2 %	23	2.1 %	27.2 %	14.4	17B	16.0 %	18	27.0 %	8	8.3 %	42
Minnesota	29.1 %	24	3.0 %	26.0 %	8.3	40	9.9 %	48	19.5 %	46	7.9 %	46B
Kansas	28.6 %	25	3.3 %	25.2 %	14.4	17A	11.4 %	40	21.1 %	40	10.5 %	26B
Georgia	28.3 %	26	0.8 %	27.6 %	15.8	13	18.7 %	14	23.7 %	16A	12.5 %	18A
Louisiana	27.3 %	27	0.7 %	26.7 %	21.3	5	20.3 %	8	23.6 %	18A	18.5 %	2
Virginia	27.0 %	28	0.9 %	26.1 %	12.5	19	15.4 %	20	21.2 %	39	8.1 %	43C
Michigan	25.3 %	29	0.9 %	24.4 %	14.9	16	14.2 %	29	25.1 %	12	10.2 %	29B
New Mexico	25.0 %	30	0.9 %	24.1 %	8.5	37	19.1 %	12	22.5 %	27B	19.3 %	1
Pennsylvania	23.0 %	31	0.7 %	22.3 %	15.1	15	15.1 %	23B	23.2 %	24	9.8 %	34C
Ohio	22.7 %	32	1.3 %	21.3 %	11.7	23	13.4 %	34	27.6 %	3	11.1 %	22A
Oregon	21.3 %	33	1.9 %	19.4 %	9.4	30C	13.2 %	35	21.5 %	37B	12.9 %	16
Delaware	20.0 %	34	0.6 %	19.4 %	11.6	24	14.4 %	28	25.4 %	9	9.9 %	33
Washington	18.0 %	35	0.8 %	17.2 %	9.4	30A	8.1 %	50	22.4 %	30D	9.5 %	37
Texas	17.5 %	36	0.9 %	16.6 %	11.4	25	21.3 %	5	22.4 %	30C	14.9 %	7
Colorado	15.5 %	37	1.1 %	14.4 %	5.3	48	10.4 %	45	22.5 %	27A	8.5 %	40

Table 1. Percentage of Rural Population, Fire Death Rates, and Other Explanatory Characteristics, by State (Continued)

| State | Rural Population in 2000 | | | | Fire Deaths per Million Population 1997-2001 | | Education — Adults without 12 Years of School Average of 1998 and 2000 | | Smoking — Adults Who Are Current Smokers 1999 Percentage | | Poverty — People below Poverty Line, Average of 1998 to 2000 | |
	Rural Population Percent	Percent Rural Population Rank	Farm Dwellers Percent	Nonfarm Rural Percent	Average	Rank	Percent	Rank	Percent	Rank	Percent	Rank
Maryland	13.9 %	38	0.5 %	13.5 %	10.9	26	14.8 %	25	20.3 %	43	7.3 %	50
New York	12.5 %	39	0.3 %	12.2 %	10.7	27	18.0 %	15	21.9 %	35	14.7 %	8B
Connecticut	12.3 %	40	0.1 %	12.2 %	8.1	43	14.1 %	30A	22.8 %	26	7.7 %	48
Illinois	12.2 %	41	1.0 %	11.1 %	12.2	21	15.2 %	22	24.2 %	14	10.5 %	26A
Arizona	11.8 %	42	0.2 %	11.7 %	8.4	38A	16.5 %	17	20.0 %	45	13.5 %	12
Utah	11.7 %	43	0.6 %	11.2 %	4.2	50	10.0 %	46A	13.9 %	50	8.1 %	43B
Florida	10.7 %	44	0.2 %	10.4 %	8.2	41A	17.1 %	16	20.7 %	41A	12.0 %	20A
Rhode Island	9.1 %	45	0.1 %	8.9 %	7.9	44A	19.0 %	13	22.4 %	30B	10.2 %	29C
Massachusetts	8.6 %	46	0.1 %	8.5 %	8.7	34A	14.7 %	26A	19.4 %	47	10.2 %	29A
Hawaii	8.4 %	47	0.5 %	8.0 %	5.1	49	14.0 %	32A	18.6 %	49	10.6 %	25
Nevada	8.4 %	48	0.2 %	8.2 %	8.7	34C	14.1 %	30B	31.5 %	1	10.1 %	32
New Jersey	5.7 %	49	0.1 %	5.5 %	8.7	34B	13.1 %	36	20.7 %	41B	8.1 %	43A
California	5.5 %	50	0.3 %	5.2 %	6.4	47	19.4 %	10	18.7 %	48	14.0 %	11

Notes: This table shows the percentage of rural residents in each State in 2000, according to U.S. Census data, the rank of each State in terms of percent population living in rural areas, the percent of State population living on farms, and the percentage living in nonfarm rural areas of each State in 2000. The annual average death rate per million population for 1997 to 2001 is shown for each State. Letters are used in the "Rank" columns to indicate ties and are assigned to States based on alphabetical order. For example, Alabama and New York both have poverty percentages of 14.7 percent, and so they are listed with ranks 8A and 8B, respectively.

Sources: NE-MW Economic Data from Northeast-Midwest Institute calculations based on data from U.S. Department of Commerce, Census Bureau, 2000 Census, Summary File 3, Table P.5 Urban and Rural, data extracted via *http://factfinder.census.gov/*, accessed online at *http://www.nemw.org/poprural.htm* on May 23, 2005.

Table 1. Percentage of Rural Population, Fire Death Rates, and Other Explanatory Characteristics, by State (Continued)

Sources (continued):

Hall, John R. Jr. *U.S. Fire Death Rates by State.* Quincy: National Fire Protection Association. October 2004, "Table 3. 1997-2001 Rates of Death Due to Fire, Flames or Smoke per Million Population by State and Potentially Explanatory State Characteristics." pp. 9-12.

Hall's source notes State "National Center for Health Statistics mortality data sorted by International Classification of Diseases codes, as sorted and analyzed by U.S. Consumer Product Safety Commission (1980-1998) and National Safety Council (1999-2001). Deaths included are those coded E890-E899 (1980-1998) and X00-X09 (1999-2001). Figures do not include codes F63.1 (pathological fire-setting) and W39 (fireworks discharge), which would add less than 1 percent to the total each year. Figures do not include codes X76 and X97 (suicide or homicide by smoke, fire, or flames), which would add about 8-10 percent to the total each year. These four codes are included in State-by-State analyses by the U.S. Fire Administration. Figures do not include fire deaths in vehicles, which would add about 20 percent to the expanded total (with X76 and X97) each year; those deaths cannot be readily isolated in NCHS mortality data. Also, State resident population figures are taken from the *Statistical Abstract of the United States.* Education, smoking, and poverty figures are taken, respectively, from the *Statistical Abstract of the United States 2003* (Table 231), *2001* (Table 192), and *2002* (Table 673)."

The Rural Population

The U.S. Census Bureau considers areas other than urbanized areas or urban clusters to be rural. An urbanized area has a nucleus (may or not be a unique city) with at least 50,000 residents. Such an area also has a core of at least one contiguous block group of less than 2 square miles with 1,000 people per square mile. Urban clusters have similar cores, but they have populations of from 2,500 to 49,999.

Metropolitan statistical areas, as defined by the Office of Management and Budget, include "central or core counties with one or more urbanized areas, and outlying counties that are economically tied to the core counties as measured by work commuting." Micropolitan statistical areas include a) nonmetropolitan counties with at least one urban cluster of 10,000 or more residents, and b) noncore counties that lack these urban clusters. Both types of nonmetropolitan counties often are included in studies of rural conditions. [1]

The American Housing Survey (AHS) used the 1980 U.S. Census definitions for urbanized areas (incorporated places and densely settled surrounding areas [at least 1.6 people per acre] with a combined population of at least 50,000). "Other urban areas" are areas with populations of at least 2,500 that are not inside the urbanized areas. Rural housing included that which was "not classified as urban." [2]

Cushing Dolbeare included housing in areas specifically classified as rural and "other urban" in discussions of rural housing. In 1995, according to these definitions, 37.2 million (38 percent) of the Nation's 97.7 million housing units were rural. Forty-four percent of the rural units were in metropolitan areas, and the remainder were outside metropolitan areas. Metropolitan areas, outside of New England, are defined in terms of whole counties.

Seventy-six percent of rural householders owned their own home, compared to 58 percent of urban dwellers. Three-quarters of rural housing units were single-family homes. Eleven percent of rural householders live in manufacturing housing; only 2 percent of urban householders live in these properties. More than one-third of urban housing was in buildings with at least two units. Rural householders tended to be older, poorer, and more likely to married and white than their urban counterparts. Housing in rural areas tends to be larger and less expensive than urban housing. Rural renters face a higher housing cost burden than homeowners.

Inadequate housing is a bigger problem in rural households (7 percent) and homes in central cities (8 percent) than in the suburbs (4 percent). Five percent of rural African-American households, 3 percent of rural Hispanic households, and 2 percent of rural white households lived in severely inadequate housing. Seventeen percent of rural African-American households, 10 percent of rural Hispanic households and 4 percent of white households lived in housing that was considered moderately inadequate.

Twenty-seven percent of all rural householders were at least 65 years old. Twenty-one percent of these older rural householders had incomes below the poverty line, and 31 percent had incomes between 100 percent and 200 percent of the poverty level.

Single parents accounted for one in ten rural householders. Forty-eight percent of the single parents were renters, and 43 percent had one or more children under six years old. Thirty-four percent of the rural single parent households had incomes below the poverty line, and 29 percent had incomes between 100 percent and 200 percent of the poverty level. [6]

Lynn Whitener referenced Atkinson's findings that rural families were more likely to use relatives for child care, that rural mothers have less education than urban mothers, have more children, and are more likely to be employed in clerical work than are urban mothers. Rural poverty has increased with changes to the structure of the family. [7]

Characteristics of a number of different rural populations and circumstances will be examined in the following section.

Older adults in rural U.S.

Joseph Belden reports that in 1995, 13.9 percent of the nonmetropolitan population were at least 65, compared to 11.9 percent of the population overall. Poverty and financial difficulties among older adults were more pronounced in these areas. Forty-six percent of the nonmetropolitan African-Americans who were at least 65 lived in poverty. Thirty-three percent of the nonmetropolitan Hispanic elders lived in poverty. Nonmetropolitan areas had larger shares of older adults with incomes under 200 percent of the poverty level.

Compared to elders elsewhere in the country, homes of older adults in nonmetropolitan areas were more likely to be owner-occupied and to have problems. The nonmetropolitan older adults were more likely to live in manufactured housing than their metropolitan counterparts and to have fewer economic resources. In 1995, more than one million nonmetropolitan housing units occupied by older adults lacked adequate heating equipment. Almost half a million (468,000) nonmetropolitan elder-occupied units had severe or moderate problems with heating, plumbing, electrical systems, maintenance, kitchens, and/or hallways.

Rental housing is less available in nonmetropolitan and rural areas, which can be a challenge for individuals who no longer can cope with the maintenance issues associated with home ownership. [8]

Older adults in rural parts of the U.S. are more likely to be obese, to smoke, and to be physically inactive. Using 1993-1998 data extracted from the Centers for Disease Control and Prevention's Behavioral Risk Factor Surveillance System (BRFSS), Kumar, et al., found that 18.9 percent of rural Americans 60 and over were obese, 43.1 percent were physically sedentary and 14.6 percent were current smokers. For urban Americans of the same age, 17.6 percent were obese, 37.4 percent were physically inactive, and 13.1 percent were current smokers. [9]

The rural South

McCray reports that poor housing quality was identified as a problem in the rural South in the 1940s and has remained a concern since that time. Three-quarters of the substandard housing units in the 1980s were in the South. In 1995, the 9 percent of the Nation's housing units in the nonmetropolitan South accounted for 21 percent of U.S. occupied units with moderate physical problems, 11 percent with severe problems, and 12 percent of the households with income below the poverty level. Thirty percent of the people in the 16 southern and border States lived in rural areas. According to the 1990 census, 22 percent of the rural southern population had incomes below the poverty line.

Mississippi, Texas, and Kentucky had the largest shares of substandard rural housing, with Mississippi, Texas, and Louisiana having the largest share of total substandard housing. Housing problems tend to be greater in the 214 counties of the Lower Mississippi Delta (LMD). The LMD includes counties in Mississippi, Arkansas, Louisiana, Tennessee, Kentucky, Missouri, and Illinois. Larger shares of owner-occupied housing in the LMD had values under $15,000 as compared to other counties in those States. In the LMD region, the infrastructure in poor and minority neighborhoods was inferior to that found in other communities. [10]

Wimberley and Morris describe the Southern Black Belt as a crescent-shaped belt of counties in Virginia, North and South Carolina, Georgia, Florida, Alabama, Mississippi, Tennessee, Arkansas, Louisiana, and Texas that have more than the average percentage of African-American residents. Although agriculture is an important part of the region's economy, the number of Southern farms has declined. Very few farms are operated by African-Americans. Forty-four percent of the rural population (45 percent of nonmetropolitan population) in the U.S. lives in the South. Seventy-nine percent of the U.S. nonmetropolitan African-American population lives in the Black Belt. Ninety-one percent of all nonmetropolitan or rural African-Americans live in the South.

The 11 States in the Black Belt are over-represented in lists of highest unemployment, poverty, infant mortality, and hunger. The authors note that "The 11 Black Belt States contain 35 % of the Nation's poor, 43 % of the nonmetropolitan poor, 51 % of the African-American poor, and 90 % of the nonmetropolitan African-American poor. Within these 11 States, the 623 Black Belt counties claim 23 % of all U.S. poverty, 28 % of the nonmetropolitan poverty, 47 % of the African-American poverty, and 84 % of the nonmetropolitan African-American poverty." (pp. 8-10)   Although white poverty in this region is less prevalent, high white poverty counties are scattered along the Louisiana Delta and lower portion of the Black Belt. Areas of the country with high levels of white poverty (the Appalachians, the Missouri-Arkansas-Oklahoma area, southwest Central, and upper West) tend not to have high levels of African-American poverty.

Forty percent of the U.S. residents who lack high school diplomas live in the South, with 21 percent in the Black Belt. The majority of counties in the Black Belt and the rest of Florida are in the top quartile in percent of African-Americans without high school diplomas. Counties in the top quartile of whites without diplomas are more common in the upper South from the Appalachian Mountains to Oklahoma. A strong correlation exists between the lack of education and poverty. [11]

Harris and Zimmerman reported that in 2001, 16.3 percent of U.S. children lived in poverty. The poverty rate was higher (20.3 percent) in nonmetropolitan areas than in metropolitan areas (15.4 percent). In the South, 18.9 percent of all children lived in poverty, with 17.3 percent of the metropolitan children and 24.9 percent of the nonmetropolitan children living under these conditions.

Child poverty rates also vary by race and ethnicity. Nationally, 40.5 percent of African-American children, 17.0 percent of white children, and 32.3 percent of Hispanic children living in nonmetropolitan areas of the country were living in poverty. Interestingly, the South had slightly lower percentages of African-American (39.8 percent) and Hispanic (31.9 percent) nonmetropolitan child poverty, while the 19.8 percent white nonmetropolitan child poverty was slightly higher.

Investment in education and training has been weak in the rural South, leaving a work force with fewer skills. With many part-time and seasonal positions, working poverty is common. Parents of poor children in the rural South are younger and less educated than parents of poor children elsewhere. Fifty percent of rural Southern children in mother-only families lived in poverty, compared to 16 percent in two-parent families. Family income influences housing quality, neighborhood, and educational and social opportunities. Welfare reform has led to many families leaving welfare rolls, but remaining in poverty. Resources often are lacking for child care, transportation, and prevention programs in rural areas. Racial and sexual discrimination has been a factor in excluding minorities and women from educational and employment opportunities that would enable an adequate, stable income. Families lack the resources to break the cycle. [12]

Lynn Whitener referenced Stoneman, et al., study of 90 African-American children between 9 and 12 years old. The children's parents were married, and the families lived in the rural South. More conflict, more depression, and less support were found in families with fewer financial resources. Conflict and "loss of family optimism" were associated with a lack of youth self-control or self-discipline and diminishing achievement. [7]

Appalachia

Mark Mather found that 75.5 percent of the nonmetropolitan Appalachian housing units were owner occupied in 2000. Home ownership was highest (76.9 percent owner-occupied) in distressed counties and lowest in the attainment counties (69.2 percent). (Attainment counties have income, unemployment, and poverty rates that are equal to or better than the national average.) Manufactured homes accounted for 19.5 percent of the housing units in nonmetropolitan Appalachia, 24.7 percent of the homes in distressed Appalachian counties, and only 4.4 percent of the attainment Appalachian counties. The lower cost of manufactured housing has contributed substantially to the above-average levels of home ownership in the region. Nationally, 66.2 percent of U.S. homes were owner-occupied in 2000.

In 2000, 2.4 percent of all U.S. households had no access to telephones, but 4.7 percent of the nonmetropolitan Appalachian households lacked this access, and the percentage lacking telephone access in distressed parts of Appalachia was 6.9 percent. In 23

Appalachian counties, 10 percent or more the households did not have access to telephones in 2000. The data do not indicate whether the problem is lack of funds or lack of available phone service in the community. [13]

The rural Midwest

Krofta, Cull, and Cook report that out-migration from the rural Midwest in recent years has left an aging population and an aging housing stock. In the Midwest and the rest of the country, nonmetropolitan households are more likely to be white, older, and married. In 1995, 28 percent of the nonmetropolitan Midwest heads of households were at least 65 years of age as compared to 21 percent nationally. Housing costs are lower in the rural Midwest than in the country as a whole. One-family, detached dwelling units accounted for 79 percent of the nonmetropolitan Midwest housing compared to 68 percent of the Nation's housing supply. Manufactured homes accounted for 8 percent of the nonmetropolitan Midwest homes, and 6 percent of the country's housing. Twenty-five percent of nonmetropolitan Midwest housing is at least 50 years old; much of this housing needs updating.

Slightly more than half of nonmetropolitan Midwest homes use piped gas as their main source of heat. Two-thirds were getting water and sewer services from a water company or public system.

Consistent with other areas of the country, households of older adults and those headed by women and/or minorities were most likely to have inadequate housing and to use larger shares of their income for housing. Native Americans and migrant farm workers in the region are confronted with especially severe housing conditions and discrimination. Migrant workers face the added challenge of frequent moves. [14]

Migrant workers and the colonias

Slesinger and Ofstead report that 159,000, or roughly 6 percent, of paid farm workers were migrants in 1985. Their study compares the characteristics identified by interviews with migrant Wisconsin workers conducted in the summers of 1978 and 1989. The workers tended to work in either the fields or the canneries.

In 1989, 72 percent of the migrant workers in the area were male. Sixteen percent of the migrant men and 19 percent of the migrant women were functionally illiterate. Only 8 percent of the men and 14 percent of women migrant workers 25 years of age or older had completed high school compared with 76 percent of the U.S. population of that age. Sixty-two percent of the 1989 sample were married; most married couples had children or other relatives as part of their household. Three generations were present in 11 percent of the households. Only 13 percent of the migrant workers described their heath as excellent, compared to 40 percent of the U.S. population as a whole.

While all of the migrant workers in the study lived away from home in July and August, roughly 90 percent were back in their home States in the winter months. Almost 60 percent were unemployed in the winter.

The median household income for these workers (average household of 5.2 persons) in 1988 was less than half the Federal poverty level and about a fifth of the national household median income. Migrant work was the sole income source for 44 percent percent of the households. During the peak season, workers often worked double shifts 7 days a week.

Most of the agricultural migrant workers lived in employer-provided housing. One-third of the housing units did not have indoor plumbing in 1989. In these cases, separate bath houses were provided. [15]

Susan Peck reports that the majority of farm workers in the U.S. are now Latino and that many of the workers are undocumented. In the 1980s, farm workers in Delaware, Maryland, and Virginia were usually African-American or Caribbean, but as of 1992, 84 percent were Hispanic. African-American farm workers tend to be single men; the Mexican and Mexican-American workers tend to travel as families. Many of these workers have no reading skills in any language.

In areas with long growing seasons, such as Oregon, farm workers often decide to stay, and the communities become more Hispanic. It is estimated that about 700,000 farm workers are hired annually in California; 92 percent are foreign born. Just 9 percent were not authorized to work in the U.S. A growing number of farm workers are from the indigenous peoples in Mexico and Central America. Many of these people do not speak Spanish or English.

A 1993 report found that the median personal income for fieldworkers in California was between $5,000 and $7,500. Only 11 percent received food stamps, 2 percent were getting Aid to Families with Dependent Children, and 3 percent received housing assistance.

Increased enforcement of health, safety, and housing requirements and a corresponding increase in associated penalties have coincided with a decline in farmers providing housing for their workers. In 1968, 5,000 labor camps in California were licensed by the State. In 1994, there were only 1,000. Nonemployer organizations are providing more and more of this housing. Labor contracting is becoming more common and direct hiring less so. [16]

Martinez, Kamaski, and Dabir describe the circumstances of the colonias. A colonia, according to the definition used by the 1990 National Affordable Housing Authority Act is "an identifiable community in Arizona, California, New Mexico, or Texas within 150 miles of the U.S.-Mexico border, lacking decent water and sewage systems and decent housing, and in existence as a colonia before November 28, 1990." (p. 50) Other agencies and jurisdictions use different definitions. In 1995, the Texas Water Development Board estimated that roughly 280,000 people lived in 1,193 colonias in Texas, with 60 percent of this population in the four counties of the Lower Rio Grande Valley, counties that rank among the most impoverished in the country. Estimates of the number of New Mexico colonias and residents vary widely, ranging from 15 to 60 colonias, with 14,600 to 100,000 residents. In 1987, a Congressional Research Service (CRS) study found 25,000 in colonias of San Diego County and 11,500 in Imperial County colonias. Many rural Latino communities strongly resemble colonias but are outside of the border area. The 1987 CSR study found 50 to 55 colonias in Arizona.

Colonias do not fit neatly into traditional definitions of rural versus urban, are limited to four States, and are often ineligible for programs. Because colonias are physically, and generally legally, isolated, basic infrastructure such as water, sewer, and paving lack economies of scale. Many colonias are not in cities and colonias' residents do not have sufficient income for user fees or many taxes. Colonia residents lack political power. They comprise only a small share of the local population in any voting district. Although advocacy by and services from community organizations were common in the colonias during the 1960s and 1970s, funding cuts had a major impact on these activities. [17]

The rural homeless

Whitener referenced a 1994 study by First, Rife, and Toomey of nonurban and rural Ohio homeless. The authors found a larger share of young women with children who became homeless due to family conflict than had been seen in national studies. They also found that homeless, single-parent rural families were more likely to rely on relatives for resources than comparable families in urban areas who had greater access to social services. [6]

Mary Stover also discussed the problems of the homeless in the rural U.S. Many of the factors seen in rural homelessness are consistent with what is seen in urban areas: poverty, lack of affordable housing, unemployment, underemployment, substance abuse, physical or mental health problems, and, for women and children, domestic violence. Domestic violence is believed to be a larger part of the rural homeless problem. A lack of public transportation to services and work reduces options in rural areas.

Compared to the urban homeless, the rural homeless population contains more women and families, fewer men, and tends to be younger. Rural areas have fewer minorities than urban areas, and consequently fewer homeless minorities. However, rural minorities face a higher risk of homelessness than do whites. Native Americans and migrants are at higher risk. [18]

Drugs in rural areas

In 2000, Columbia University published a study of substance abuse in midsize cities and rural America. Rural areas frequently lack enforcement and treatment resources found in more populated areas. Rural eighth graders were more likely to have used marijuana (11.6 percent), amphetamines (5.1 percent), cigarettes (26.1 percent) and alcohol (28.1 percent) in the past month than were students in small (marijuana—9.4 percent, amphetamines—3.1 percent, cigarettes—16.0 percent, alcohol - 23.4 percent) and large metropolitan areas (marijuana—8.6 percent, amphetamines—2.5 percent, cigarettes—12.7 percent, alcohol—21.7 percent). Use of cocaine and amphetamines by rural tenth and twelfth graders was higher than use by students in large urban areas.

Among adults, marijuana use is less common in rural areas (past month use from 1997 to 1998 of 9.5 percent rural versus 15.8 percent small metro and 13.0 percent large metro use among 18-25-year-olds), but no statistically significant differences were seen in use of other illicit drugs. From 1997 to 1998, 47.7 percent of rural young adults (18-25) had smoked a cigarette in the previous month compared to 36.7 percent in large metropolitan areas and 42.1 percent in small metropolitan areas. In 1992, Native Americans were found to have much higher rates of illicit drug use, alcohol consumption, and smoking. The authors reference Beauvais and Segal's 1992 paper "Drug Use Patterns among American Indian and Alaska Native Youth: Special Rural Populations." The Columbia University authors note that "Youths on the reservation were 3.5 times likelier to have tried marijuana, 5.8 times likelier to have tried stimulants and 8.3 times more likely to have tired heroin than were youths in a nationwide sample."

Although no significant differences were seen among adult rural or urban methamphetamine users, rural 10th and 12th graders were substantially more likely to have used methamphetamines within the past year. Methamphetamine laboratories tend to be located in areas with low populations so that the fumes don't attract attention. Initially found more in the West, they are becoming increasingly common in the Midwest, with drug operatives installed among rural laborers. The manufacturing process is dangerous, with signs of explosions frequently found in the labs. [19]

The Rural Fire Problem

In 1997, the USFA issued its report on *The Rural Fire Problem in the United States.* This report used 1993 to 1995 NFIRS data and 1983 to 1988 mortality data from the National Center for Health Statistics to examine and contrast fires and fire deaths occurring in rural and nonrural areas of the U.S.

Definition of "rural" used in USFA report

The USDA's Rural-Urban Continuum, or Beale Codes, was used to define rural for this analysis. Areas with populations of 2,500 or fewer or up to 19,999 but not adjacent to a metropolitan area were considered rural in this analysis. In 1993, according to this definition, 19.4 million (7.5 percent of the population) lived in the 45.7 percent of U.S. counties that were considered rural.

Incident types in rural and nonrural areas

From 1993 to 1995, the incident types for reported fires and civilian fire deaths were similar in rural areas and in the U.S. as a whole. Outside fires accounted for 45 percent of fires reported in rural areas and 43 percent of the reported fires in the entire Nation. Residential structure fires accounted for 25 percent of rural fires. Residential structure fires caused 69 percent of the rural civilian fire deaths and 60 percent of the rural civilian fire injuries. In the U.S. as a whole, residential structure fires accounted for 23 percent of the reported fires, 72 percent of the civilian fire deaths and 68 percent of the civilian fire injuries. (NFPA historically has found a larger share of fire deaths in residential occupancies. The NFIRS database historically has shown a larger share of vehicle fire deaths than NFPA's survey. NFPA contacts fire departments to confirm that reported vehicle fire deaths resulted from the fire, not trauma.)

The graphs showing 1993 to 1995 statistics reflect the data found in the USFA's report.

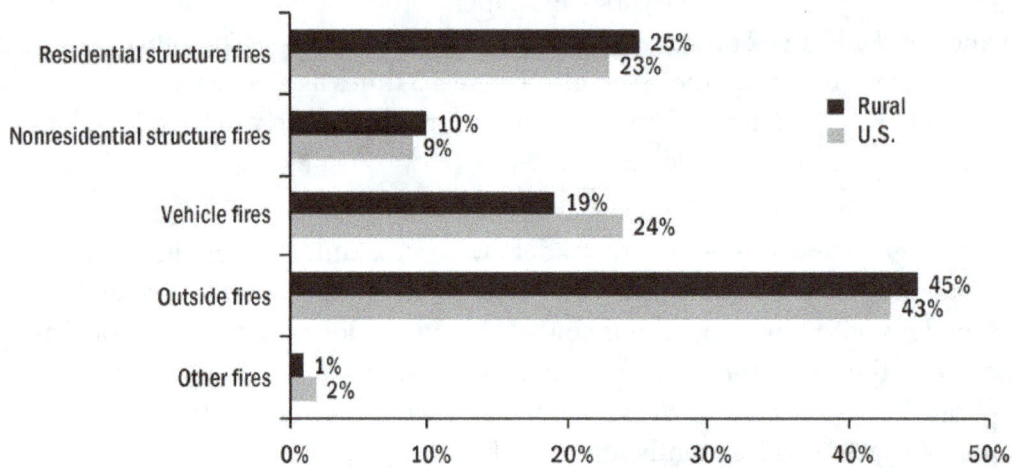

Reported Fires in Rural Areas and Entire U.S.
by Incident Type: 1993-1995

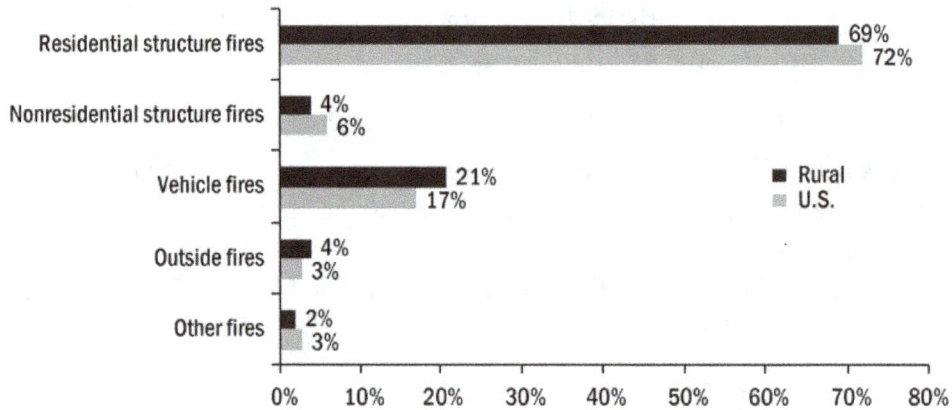

Civilian Fire Deaths in Rural Areas and Entire U.S.
by Incident Type: 1993-1995

Cause profile is different in rural areas.

Forty-five percent of the rural outside fires were caused by open flame, 16 percent by arson, and 9 percent by natural causes. Arson caused 44 percent of the nonrural outside fires.

Regional differences in outside fires

Forty-nine percent of the South's rural fires occurred outside, compared to 43 percent in the North. When divided by East and West, outside fires accounted for 55 percent of rural fires in the West, but only 36 percent in the East. Although open flame caused more than 40 percent of the rural outside fires in both the East and West, arson caused 29 percent of the eastern outside rural fires, compared to 12 percent of the western rural fires.

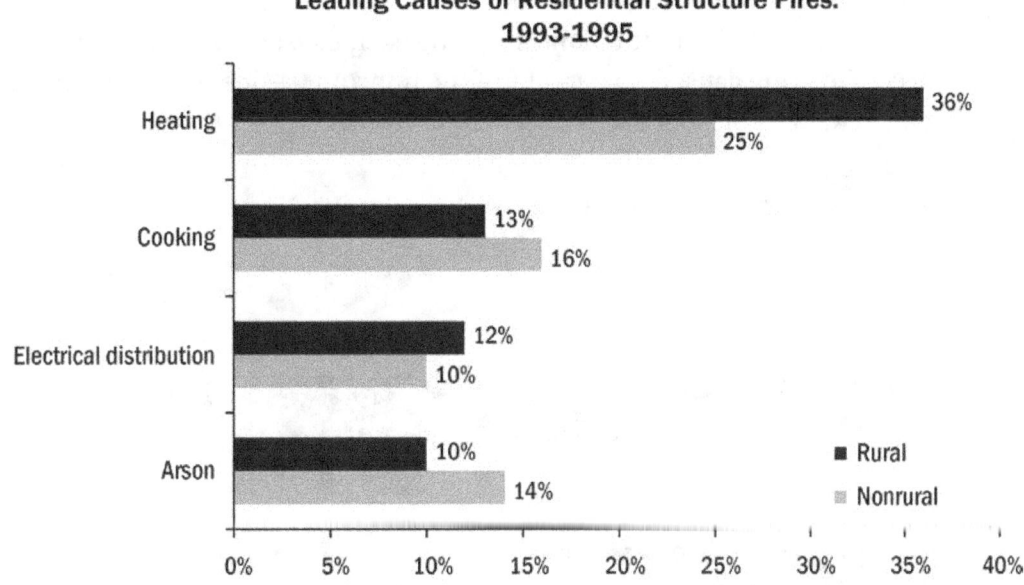

Leading Causes of Residential Structure Fires: 1993-1995

Rural residential structure fires

Rural residential fires were more likely to be caused by heating equipment, to occur in properties that had no smoke alarms at all, and to have flame damage extend to the entire structure. Thirty-six percent of the rural residential fires were caused by heating, 13 percent by cooking, and 12 percent by electrical distribution equipment. Twenty-six percent of the fatal residential rural fires were caused by heating, 23 percent by smoking, and 17 percent by electrical distribution equipment. Smoking caused 28 percent of the nonrural, fatal residential fires; 17 percent were arson, and heating caused 12 percent.

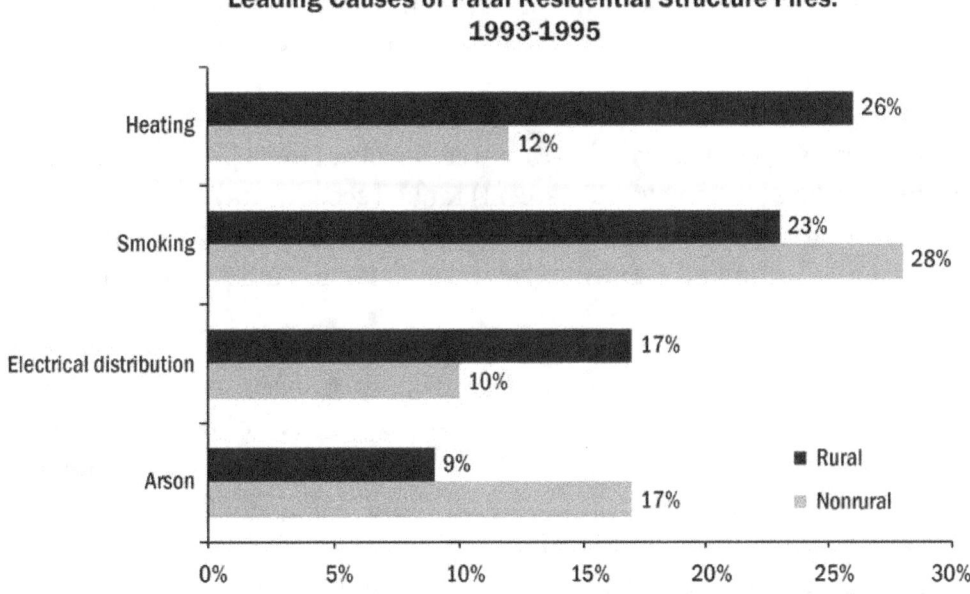

Leading Causes of Fatal Residential Structure Fires: 1993-1995

Heating and cooking each caused 23 percent of the rural residential fires with injuries. Children playing, electrical distribution and smoking each caused 10 percent of these fires with injuries. Cooking caused 30 percent of the nonrural residential fires with injuries; smoking and children playing each caused 12 percent.

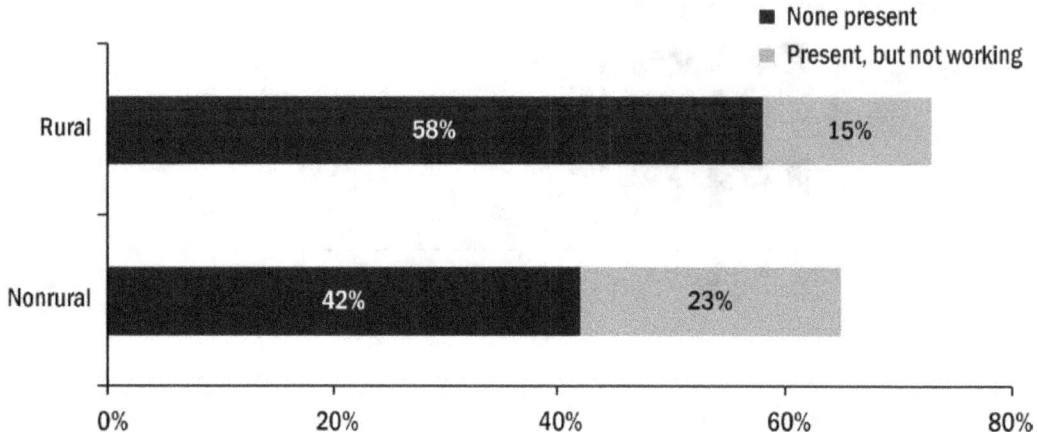

Residential Fires with No Working Smoke Alarms: 1993-1995

Smoke alarms

Almost three-quarters (73 percent) of rural residential fires occurred in properties without working smoke alarms compared to 65 percent in nonrural properties. The larger difference seen was in presence versus nonoperating. In 58 percent of the rural residential fires, no smoke alarms were present at all; in 15 percent, smoke alarms were present but not operating. In 42 percent of the nonrural incidents, no smoke alarms were present at all; in 23 percent, these devices were present but not operating.

Flame damage extended to the entire structure in 29 percent of the rural residential structure fires but only 17 percent of such incidents in nonrural areas.

Rural heating equipment fires in USFA report

Fixed area heaters, including wood stoves, were involved in 38 percent of the rural residential heating fires. Chimneys (25 percent) ranked second, and fireplaces (11 percent) ranked third. Adhesive, resin, or tar was the type of material first ignited in nearly half of the rural residential heating fires. Sawn wood was first ignited in 19 percent of these fires.

Regional differences in rural residential fires

Greater differences were seen in residential structure fire causes between North and South than between East and West. Heating caused 39 percent of the rural residential structure fires in the North, but 29 percent in the South. Electrical distribution fires equipment caused 12 percent of these fires in both the North and the South. Cooking caused 11 percent of the rural residential fires in the North, and 18 percent in the South. Arson caused 8 percent in the North and 12 percent in the South.

**Equipment Involved in the Ignition
of Rural Residential Heating Fires: 1993-1995**

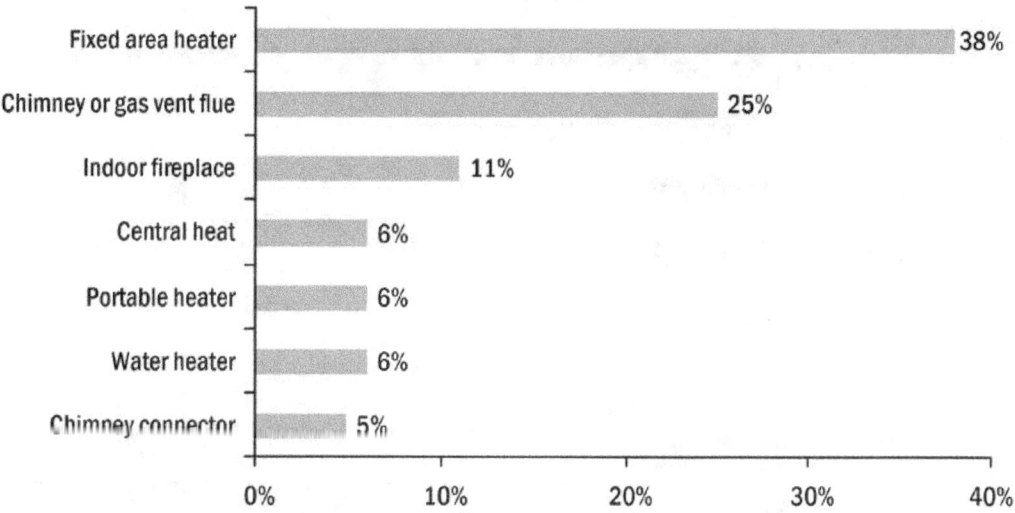

Smoke alarms were present and operated in 30 percent of the residential structure fires in the North, but only 21 percent of such incidents in the South.

The percentage of rural residential fires originating in the chimney was three times as high in the North as the South. The kitchen was the leading area of origin in the South and the second most frequent area in the North. Bedrooms were the second most frequent area of origin in the South.

**Leading Causes of Rural Residential Structure Fires
in the North and South: 1993-1995**

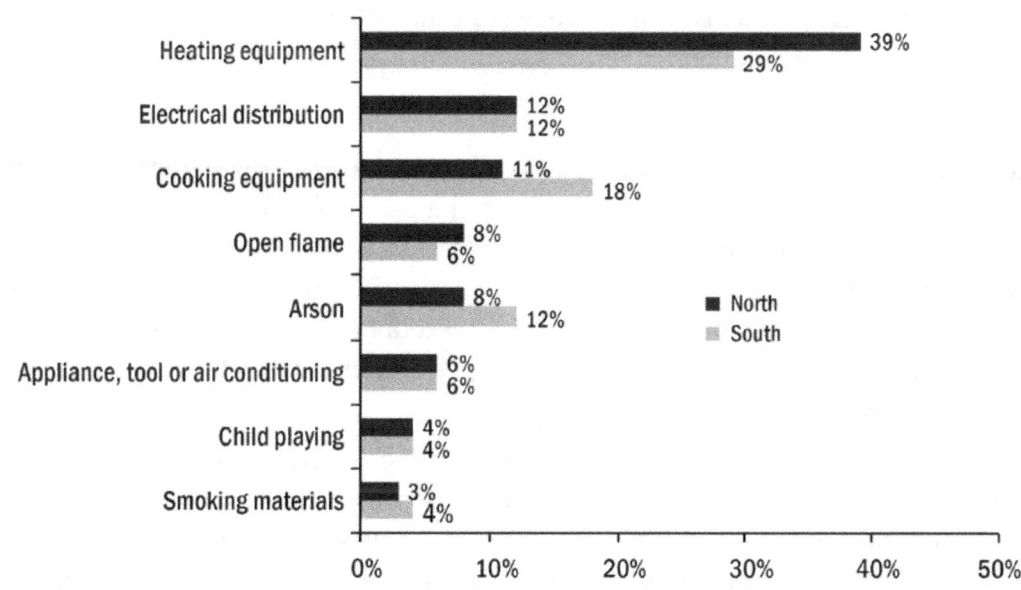

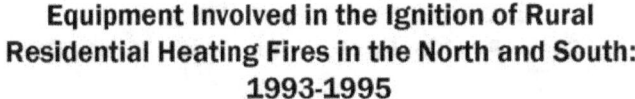

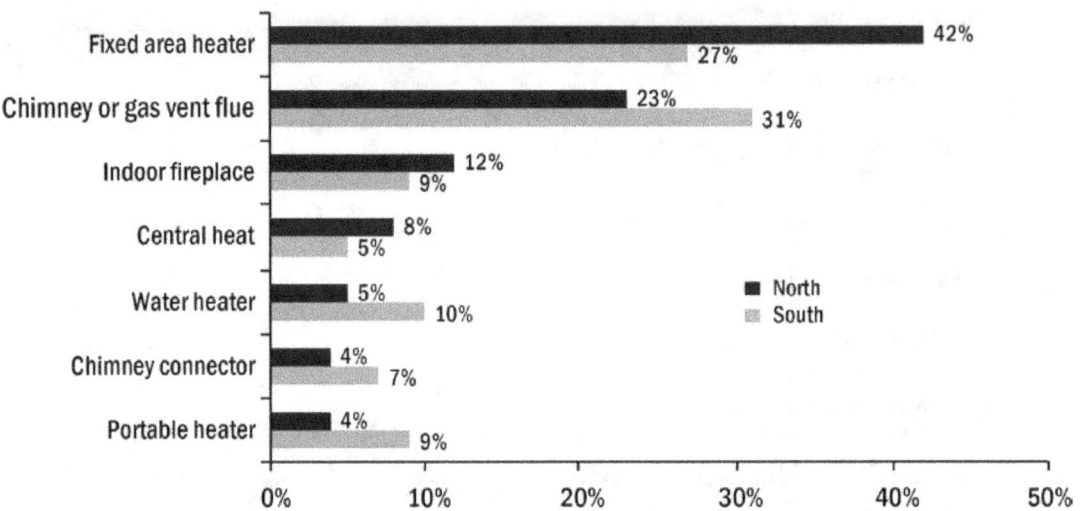

Central heating equipment caused only 8 percent of the rural residential heating equipment fires in the North, and 5 percent in the South. Fixed area heaters, including wood stoves, and chimney or gas vent flues dominated the rural fire problem, although the fixed area heaters were a bigger problem in the North, and chimney or flue fires were a bigger problem in the South.

Rural manufactured home fires

Heating caused 23 percent of the rural manufactured home fires, still the leading cause, but a smaller share of fires than in other rural housing. Electrical distribution equipment caused 19 percent and cooking 14 percent. Cooking and heating each caused 19 percent of the rural manufactured home fires in the South. In the North, heating caused 26 percent of the rural manufactured home fires and cooking 11 percent.

Rural fire deaths by race

According to mortality data from the National Center for Health Statistics, from 1983 to 1988, an average of 5,764 U.S. fire deaths occurred per year. White victims accounted for 480 of the 676 rural victims per year. Little difference is seen in the percentage of fire victims by race or gender in rural versus nonrural areas. The racial picture is different when death rates per million population are considered. From 1993 to 1988, the overall death rate for the U.S as a whole was 23.5 deaths per million. In rural areas, it was 30.9, and nonrural it was 22.8.

Rural whites had a fire death rate of 24.7, rural Native Americans had a rate of 60.7, and rural African-Americans had a rate of 88.6, the highest of any group studied. The death rate for rural Native Americans was almost three times as high as the rate for non-rural Native Americans. Rural African-Americans had a fire death rate 60 percent higher

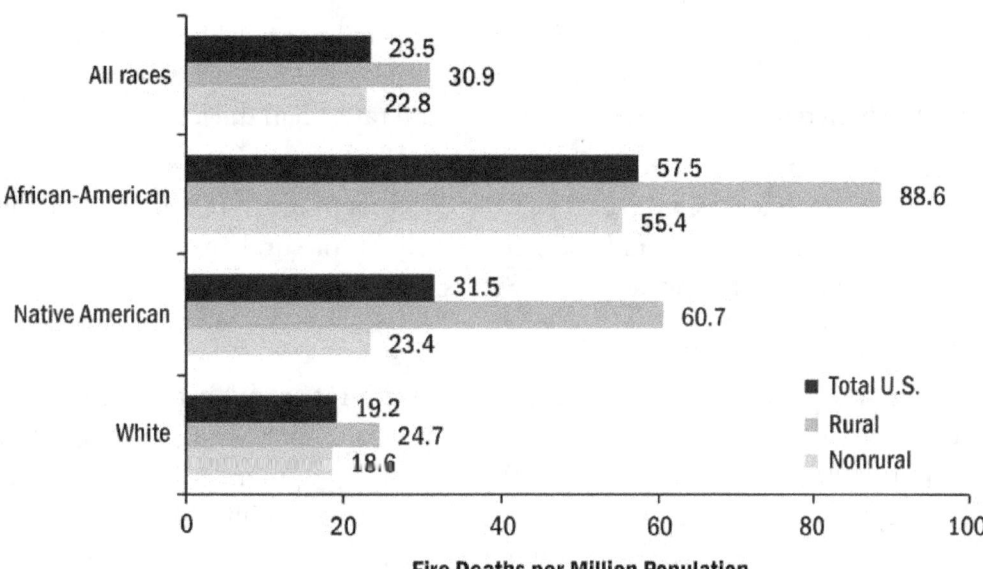

1983-1988 Fire Death Rates by Race

than their nonrural counterparts. The fire death rate among rural whites was 33 percent higher than white nonrural residents.

Rural fire victims were slightly more likely to be between 1 and 24 years old, or over 85 than nonrural victims. [20]

Gomberg and Clark's study of States with high and low fire death rates

In an older study looking at 1978 or 1978-1979 rural and nonrural civilian fire deaths in 12 States, Gomberg and Clark compared scenarios and fire death rates for high and low fire death rate States as identified by earlier data. The high death rate States included Mississippi, Alabama, Arkansas, Tennessee, Georgia, and Oklahoma. The low death

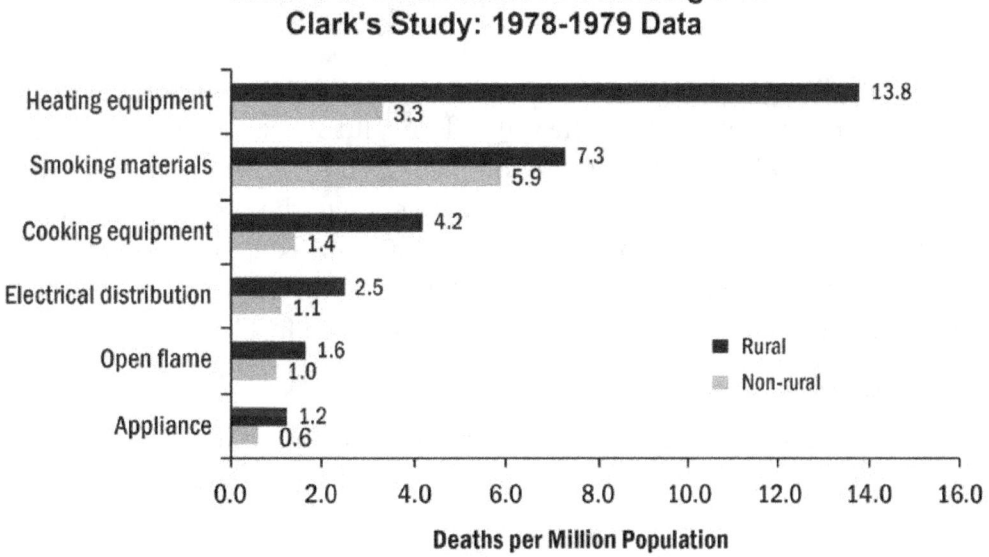

Causes of Fire Deaths in Gomberg and Clark's Study: 1978-1979 Data

rate States included Connecticut, Utah, Wisconsin, California, Florida, and Delaware. A death was considered rural if the fire occurred in an area with less than 2,500 population.

The overall rural fire death rate was 2.5 times the overall nonrural fire death rate, with rural areas having higher fire death rates in both the high and low fire death rate States. Predictably, the high death rate States had higher rural rates than the low death rate States.

Rural fires started by heating equipment had a fire death rate of 13.8 deaths per million population, four times the nonrural heating equipment death rate of 3.3. The heating equipment fire death rate was roughly 50 percent higher for rural areas in high death rate States than for rural areas in low death rate States.

Solid-fueled heating equipment fires

Solid-fueled heating equipment had a rural fire death rate roughly twice that of gas-fueled equipment. The rate for liquid-fueled heating equipment was much lower than either. The improper installation of solid-fueled heating equipment (wood stoves, fireplaces, chimneys) caused a rural death rate of 2.43 deaths per million population and a nonrural rate of 0.26. In this scenario, a wood stove may be too close to a wood wall, chimneys or vents may have inadequate clearance from wall coverings or framing, or the floor is not protected from heat or flame.

The rural rate for combustibles (furniture, linens, trash, etc.) placed too close to solid-fueled heating equipment was 0.76/million compared to a nonrural rate of 0.10/million.

Rural States had a death rate of 0.50/million for fires that started when flammable liquids were used to kindle fires in wood stoves and fireplaces. The nonrural rate was 0.02/million. Alcohol was a frequent factor in this scenario.

Worn clothing ignited by heating equipment

Ignitions of worn clothing by all types of heating equipment resulted in a rural fire death rate of 2.26/million and a nonrural rate of 0.8/million. Victims were frequently older adults who had been sitting close to a wood stove or local heater. Gas-fueled equipment was seen as a larger problem in States with high death rates; solid-fueled equipment was a bigger factor in rural areas.

Portable heater fires

Portable space heaters that ignited the walls, floors, furniture, trash, soft goods, or bedding had a rural fatality rate of 1.66/million compared to 0.48/million in nonrural areas. These fires occurred when the heater was too close to the items ignited. In many cases, portable heaters were used in place of central heat or fixed local heaters. The ignition of wood paneling by solid- and liquid-fueled heaters was a particular problem in rural areas. Fabric ignition by electric heaters was a problem in all areas.

Gas-fueled furnaces and water heaters in manufactured homes

Gas-fueled furnaces and water heaters that ignited wood paneling, walls, ceilings, floors, framing, or insulation in manufactured homes had a rural fatality rate of 0.68

deaths per million residents. None of these deaths were seen in nonrural areas of the States studied. Older manufactured housing, common in rural areas, often lacked fire-rated compartments around heating equipment. Instead, combustible paneling was used.

Smoking

Fires started by smoking materials had an overall fire death rate of 6.1, with the rural rate of 7.3 only 25 percent higher than the nonrural rate of 5.9. Smoking in bed and upholstered furniture ignitions were the most common scenarios.

Cooking

Fires started by cooking had an overall fire death rate of 1.9. The rural rate of 4.2 was twice as high as the overall and three times the nonrural rate of 1.4 deaths per million population.

As noted earlier, rural areas had a cooking fire death rate three times that of nonrural areas. In rural areas, the death rate from unattended cooking (including fires in which the individual was asleep, unconscious, or incapacitated) that ignited cooking materials or walls was 1.18 deaths per million population. In nonrural areas, the rate was 0.28. Alcohol was a factor in about 35 percent of the deaths in both rural and nonrural areas.

Rural areas had a death rate of 0.50/million from fires that occurred when gas or walls were ignited by cooking equipment with installation or maintenance deficiencies. None of these fires were seen in the nonrural areas. Leaking liquefied petroleum gas (LPG) was the most common problem.

Rural areas had a death rate of 0.67/million from fires that occurred when cooking equipment ignited misused gases or flammable liquids. The nonrural rate was 0.18/million. In some cases, flammable liquids were used for cleaning or stripping and ignited by pilot lights. In other cases, these liquids were used to kindle wood stoves. The improper fueling or lighting of gas stoves was also a common scenario.

Electrical distribution equipment

Electrical distribution equipment fires had a rural death rate of 2.5, more than twice the 1.1 nonrural rate and almost twice the 1.4 overall rate. Fixed wiring, switches, and receptacles that ignited walls, ceilings, framing, or insulation had a rural fire death rate of 1.42/million compared to 0.32/million for nonrural areas. Little difference is seen in rural and nonrural death rates from fires caused by cords, plugs, or bulbs

Open flame

Open flame caused 1.6 rural deaths per million population, compared to a nonrural rate of 1.0 and an overall death rate of 1.1. Two scenarios dominated. The death rate from outside open burning was 0.67/million in rural areas and 0.12/million in nonrural areas. The open burning was generally done for waste disposal of leaves, brush, or trash. Most victims were older men whose clothing ignited while engaged in open burning activities.

The death rate from candle fires started inside was 0.58/million in rural areas and 0.26/million in nonrural areas. The rate is highest in rural areas of high fire death States. Candles were used for light in a notable portion of these incidents. Most victims were juveniles.

Other causes

Appliances had a rural death rate of 1.2, twice the nonrural rate of 0.6. The overall rate was 0.7. The rural rate of 1.1 deaths per million population for fires started by children playing was comparable to the 1.0 rates seen in nonrural areas and overall. Rural areas had a rate of only 0.7 deaths per million population for incendiary and suspicious fires. This is roughly one-third of the nonrural rate of 2.0, and less than half the overall rate of 1.7.

Race

The nonwhite fire fatality rate was more than three times as high as that for whites in rural areas of high fire death rate States, and roughly 50 percent higher for nonwhites in low fire death rate States.

Occupancy

Sixty-eight percent of the rural fire deaths and 60 percent of the nonrural deaths resulted from fires in one- or two-family dwellings; 25 percent of the rural deaths and 8 percent of the nonrural fatalities occurred in manufactured homes; 2 percent of the rural fire deaths and 23 percent of the nonrural occurred in apartments; and 6 percent of the rural and 9 percent of the nonrural fire fatalities occurred in other residential properties.

Smoke alarms

Smoke alarms were not common during the period of this study. Ninety-four percent of the rural and nonrural deaths occurred in properties without smoke alarms.

Fire department response

Fire departments responded to 94 percent of the residential fire deaths in this study in which response status was known. In rural areas of high fire death rate States, nearly 15 percent of the deaths had not been reported to fire departments. The largest share of unreported fire deaths were caused by heating equipment fires, often clothing ignitions. Thirty-nine percent of the clothing ignitions in this study were not reported. Some rural areas lack fire department protection. [21]

Gunther studied role of climate and poverty in rural fire deaths

Paul Gunther reported on a FEMA analysis of National Center for Health Statistics (NCHS) death certificate data for fire deaths during 1974-1978 in communities with populations under 10,000. (Transportation deaths were not included.) The northern section of the U.S. had moderate to high rural fire death rates. The correlation between rural fire deaths in the North and a) heating degree days and b) freezing days was 0.71 and 0.76, respectively.

Rural fire death rates were low in the central part of the country, and high in the South (Alaska and Hawaii were excluded from the analysis.) Among the 14 central continental States above 42.5° latitude, rural fire death rates were low in three, intermediate in eight and high in three. In the rural South, fire death rates were high in all States except Florida. Rates were very high in Arkansas, Louisiana, Mississippi, Alabama, and South Carolina.

All of the Southern States had high rates of rural poverty. Missouri was the only State outside of the South to have a high rate. Rural poverty rates were either low or intermediate in the remainder. A statistical correlation of 0.79 was found between rural poverty and fire death rates for the national rural population as a whole.

Gunther garnered additional insights from Gomberg and Clark's study. NFIRS data for rural areas (populations under 10,000) of six Northern States (Maine, New York, Ohio, Oregon, Missouri, and Maryland) in 1977 showed that 38 percent of the fire deaths were caused by heating, compared to 19 percent caused by smoking. Gomberg and Clark found that in five Northern States (Connecticut, Delaware, Wisconsin, Utah and California), heating caused 26 percent of the fire deaths and smoking caused 32 percent in communities with under 10,000 population. Gomberg and Clark also found that heating caused 42 percent of the rural fire deaths and smoking 20 percent in seven Southern States (Alabama, Arkansas, Florida, Georgia, Mississippi, Oklahoma, and Tennessee.)

Types of heating equipment used by race and region

According to data from the 1976 Annual Housing Survey, warm-air furnaces were the primary heat source in 42.5 percent of rural Southern households, 22.3 percent of rural African-American households in the South, and 55.6 percent of households outside the South.

Room heaters without flues were the primary source in 15.9 percent of rural Southern households, 34.3 percent of rural African-American households in the South, and 0.6 percent of households outside the South.

Fireplaces, stoves or portable heaters were the primary heat source in 11.2 percent of rural Southern households, 20.1 percent of rural African-American households in the South, and 4.3 percent of households outside the South.

Bottled gas and wood are used as heating fuels more frequently in the South, with African-American households having the highest usage. [22]

Fire, risk, and cultural factors among Native Americans

Alisa Wolf described fire safety issues in the Navajo Nation. Some homes in the Navajo Nation are more than 2 hours away from the closest fire station and, at the time of the 1990 Census, roughly three-quarters of Navajo homes did not have telephones. Firefighters must be called individually. No mechanism exists for paging. The first Navajo fire department was funded in 1982.

Some traditional cultural beliefs conflict with firefighting responsibilities. Traditionally, a medicine man's blessing would be sought before entering a burned building.

More than half of the Navajo homes used wood for heat, according to the 1990 Census, and improperly installed wood stoves cause a substantial share of the winter fires. Some fires were started by discarded ashes that had been stored in the home. Misuse of propane and kerosene also caused problems.

The risk of fire death among Native Americans (all tribes) throughout the U.S. is two to six times as high as for nonnatives. According to a 1993 report by the Indian Health Services (IHS), fires were the leading cause of unintentional injury death in Native American homes in the U.S. According to the 1990 Census, 58 percent of the Native American residents of the Navajo reservation had incomes below the poverty level; 25 percent were unemployed.

On the reservation, older manufactured homes that predate the safety standards of 1976 are common. Often, these homes have only one exit, smaller rooms, and room linings that burn more easily. The Navajo Nation does not have a fire code. A study funded by the Indian Health Service found a large share of the smoke alarms on the Devils Lake Sioux Reservation in North Dakota had been disabled because of nuisance alarms.

The Navajo Nation Interagency Fire Safety Coalition was formed in response to a fatal fire. The Coalition produced a media campaign and conducted various forms of outreach, including talking to shoppers and meetings at chapter houses. While public educators have been using positive messages for years, this is especially important for Navajo communities. In this tradition, thinking or talking about bad things increases the likelihood they will occur. [23]

In 1992, Mobley, et al., conducted in-person interviews with 68 households in Kitsap County, Washington, that had at least one member of the Suquamish Tribe. The majority of households were on a reservation. Fifteen percent of the households lived in manufactured housing. Seventy-nine percent used electricity as either a primary or secondary heat source. Almost half used wood stoves. Ninety-six percent had at least one smoke alarm, and 95 percent of the installed smoke alarms operated when tested. Fifty-nine percent had experienced an unwanted smoke alarm activation with 84 percent of these caused by cooking.

One or more smokers lived in 59 percent of the households; one-fourth had someone who smoked in bed. Someone smoked and drank alcohol concurrently in 38 percent of the households.

Nine of the 68 households (13 percent) had experienced a home or yard fire in the previous 3 years. This translates to rate of 6.4 fires per 100 households per year. Nine percent of the households had a member burned or injured by fireworks in the previous 3 years. [24]

Kuklinski, Berger, and Weaver reported on the results of unannounced visits to 120 homes with at least one Native American in the St. Michaels District of the Devils Lake Sioux Reservation in North Dakota during 1995.

Sixty-three percent of the homes used natural gas for heat; 15 percent had a wood stove or fireplace. At least one cigarette smoker lived in 73 percent of the homes. Two-thirds had household incomes below the poverty level for a family of four.

One-third of the households had no smoke alarms at all. Eighty-six percent of the homes owned by Housing and Urban Development had the devices compared to 46 percent of the privately owned homes. Smoke alarms were found in only 57 percent of the manufactured homes and 69 percent of other single-family dwellings. Forty-six percent of the smoke alarms were battery-powered, 44 percent were electrical, and 10 percent were electrical with battery backup.

Overall, 48 percent of the smoke alarms were inoperable. Eighty-six percent had been disabled due to nuisance alarms. Battery-powered smoke alarms were more likely to be disabled due to nuisance alarms than electrical units. The three photoelectric alarms had no history of nuisance activations.

Seventy-nine percent of the households with ionization alarms experienced nuisance activations. Forty-two percent reported more than 25 nuisance activations per unit in the previous year. Cooking was blamed for 77 percent of the nuisance activations; steam from the bathroom was the culprit in 18 percent. Frying was responsible for three-quarters of the cooking nuisance alarms, while baking caused roughly one-third of the cooking activations. The frequency of cooking nuisance alarms decreased as distance from the smoke alarm to the stove increased. Using a stove fan decreased nuisance activations for smoke alarms within 20 feet of the stove, but had no impact when the smoke alarm was further away. No steam activations were reported for smoke alarms more than 10 feet away from the bathroom. [25]

Alisa Wolf noted that cultural factors play a role in Navajo fire safety. Struthers and Hodge interviewed six Ojibwe spiritual leaders or traditional healers about traditional ceremonies of sacred tobacco use, and the significance and traditional practices associated with both sacred and commercial tobacco. Sacred tobacco is seen as a vital part of the culture. Commercial cigarette smoking outside of the ritual content was recognized as destructive. However, antismoking or antitobacco messages may be perceived as attacks on the culture. [26]

Burns in rural France

The U.S. is not the only country with higher rural fire death rates. Vidal-Trecan, et al., compared circumstances and characteristics of burn victims treated at 19 of 23 French burn units from September 1991 through August 1992. Municipalities with populations under 2,000 were considered rural. Only 20 percent of the French population lives in rural areas but 34 percent of the 1,234 patients treated resided in rural areas. In rural areas, 28.3 of these burns were incurred per million population compared to 18.4 in urban areas. More specifically, the burn rate was higher for both rural men and women and for rural children, teens, and young adults than for their urban counterparts. The incidence of both occupational and everyday burns was twice as high in rural areas.

Rural burn victims tended to be older and less educated. They were more likely to be retired and/or part of a couple than burn victims from urban areas.

Greater differences were seen in the burns incurred in everyday activities than in occupational burns. Rural victims of everyday activities were more likely to be 65 or over, retired, and to have some predisposing factor than were urban victims. Thirty-five percent of everyday rural burns occurred outside, compared to 22.5 percent of urban burns. Although hot liquids were the leading cause of burns in both areas, these burns were less common in rural areas. Flames or explosions caused a larger share of burns in rural areas than in urban areas. Burns caused by open fire and barbecues were also more common in rural areas.

Rural burns were deeper and more likely to cover more than 10 percent of the body surface area. A larger share of rural burns resulted in death. Rural fatalities were less likely to have had comorbidities. [27]

Minnesota's rural fire problem

The State Fire Marshal's Office of Minnesota used the USFA's report on rural fires as a model for an analysis of rural fires in their State in 1997, 1998, and 1999. However, they limited their analyses to counties with populations less than 50,000 as opposed to the 20,000 used by the USFA.

The breakdown of fire types among outside, residential structure, nonresidential structure was fairly consistent no matter the community size, although nonresidential structure fires were more common in the rural counties. In all 3 years, the percentage of urban fire deaths in residential structure fires was higher for urban counties than rural counties, and the percentage of rural fire deaths resulting from vehicle fires was consistently higher than urban fire deaths. The patterns seen in Minnesota resemble those found in the USFA's report on rural fires.

The leading causes of fires varied based on county population. In rural counties, 18 percent of the outside fires were caused by open flames, 11 percent were arson and 4 percent were started by smoking materials. Arson caused 23 percent of the urban outside fires.

Major differences were seen in causes of rural and urban residential structure fires. Heating equipment was involved in 22 percent of the rural residential structure fires in 1999, but only 9 percent of the urban incidents. Cooking caused 11 percent of the rural residential fires, but 24 percent of the urban fires.

Thirty percent of the rural fatal structure fires were caused by furnace malfunctions, and 19 percent were caused by smoking. Forty-three percent of urban fatal residential fires were caused by smoking, and 13 percent were caused by arson.

In 1999, smoke alarms were present and operated in 44 percent of the rural residential structure fires, compared to 47 percent of the urban residential fires. No smoke alarms were present at all in 33 percent of the rural residential fires. The fire was too small to activate smoke alarms in 3 percent of the rural incidents. They were present but did not operate in 15 percent of the rural residential fires.

Flame damage was confined to the object, part of room, or the room of origin in 74 percent of the urban residential structure fires during 1999, but only 58 percent of the rural incidents.

Fireplaces or chimneys were involved in 57 percent of the rural residential heating fires and 49 percent of the urban fires. Adhesive, resin, or tar was the item first ignited in 31 percent of all rural residential heating fires; sawn wood was first ignited in 23 percent of these incidents. [28]

Fires in "predominantly rural" North Carolina

Runyan, Bangdiwala, Linzer, Sacks, and Butts compared 151 North Carolina fatal fires from 13 months in 1988 and 1989 with 283 nonfatal, nonchimney home fires with someone home at the time of the fire from the same time period. The fires were in "predominantly rural areas." Heating equipment was involved in 39 percent of the fatal and 28 percent of the nonfatal fires. Space heaters were involved in 58 percent of the fatal heating equipment fires; kerosene heaters were involved in 87 percent of the fatal space heater fires. Wood stoves or fireplaces were involved in 45 percent of the nonfatal heating equipment fires; space heaters were involved in 30 percent of the nonfatal fires. Kerosene heaters were involved in 55 percent of the nonfatal space heater fires.

Thirty-one percent of the fatal and 6 percent of the nonfatal fires were caused by smoking. Cooking caused 10 percent of the fatal fires and 23 percent of the nonfatal ones. No significant difference was seen between fatal and nonfatal fires in which a) the fire was reported by telephone, b) a 9-1-1 system was or was not present, c) the fire department was or was not comprised entirely of volunteers, and d) the response time was more or less than 5 minutes. The risk of fire death risk was higher for adults 65 or older, people impaired by alcohol or drugs, and for people with physical or mental disabilities.

The risk of fire death was higher in homes without smoke alarms. The absence of these devices had a greater impact when children were present and when no one was disabled or impaired by drugs or alcohol.

Thirty-one percent of the fatal and 21 percent of nonfatal fires occurred in manufactured housing, although manufactured housing accounted for just 11 percent of North Carolina's housing units. The authors note that "fires in mobile homes with two or fewer exits were 2.6 times more likely to be associated with fatal than nonfatal fires." (pp. 861-862)

The risk of fire fatality was higher in houses that were at least 20 years old. However, smoke alarms were more likely to be found in homes built after implementation of the State building code in 1976 requiring these devices in new construction. Smoke alarms were found more frequently in manufactured housing than in other housing units. These devices were found in 54 percent of the owner-occupied units as compared to 18 percent of the rental units. [29]

Smoke alarm status in Alabama fire deaths

In their analysis of Alabama fire deaths from 1992 to 1997, McGwin, et al., found smoke alarms were present in 42 percent of the urban fire deaths but only 21 percent of the rural ones. [30]

Burn fatalities in Maine and the U.S.

Clark, Dainiak, and Reeder examined burn fatality statistics for Maine and the U.S. from 1960 through 1996. Maine had a relative fire or flame death risk compared to the whole U.S. of 1.26 in 1961 to 1964, 1.57 in 1973 to 1976, and .95 in 1993 to 1996. (A relative risk of 1.0 means that risk is identical for the two groups.)   Burn hospitalizations in Maine also declined from 34.8 per 100,000 in the mid-1970s to 10.6 per 100,000 in the mid-1990s. The authors note that oil embargoes in 1973 and 1978 resulted in an increase in the number of Maine homes heated by wood. A burn unit was established at Maine Medical Center; burn prevention programs, including smoke alarm programs, were conducted, and schools began paying more attention to burn education in the 1980s. The authors concluded that prevention, particularly the increasing use of smoke alarms and building code improvements, played a larger role in the death rate reduction than did medical care improvements. [31]

Burn injuries in "mostly rural" Iowa

Wibbenmeyer, et al., reviewed records from 1997 to 1999 of almost 1,400 burn injuries seen in emergency rooms in ten mostly rural counties in southern Iowa. Fifteen percent of the burns were flame-related. The leading cause of flame-related burns was open fires, with three-fifths resulting from open fires used to burn trash or brush. Accelerants were involved in almost half of the burns caused by open fires, but only 2 percent of burns caused by other types of fires. Trash pick-up is not provided in many parts of the studied region. Open burning is a common means of trash and brush disposal. Open burning was the second leading cause of burns to children between 5 and 15. It is unclear how old children are when they start to perform open burning, but the authors suggest that education programs may be most effective before individuals establish their own habits and practices for open burning.

"Other burning materials" accounted for 24 percent of the flame-related burns. Work equipment such as torches, welders, or plasma cutters were involved in 44 percent of the burns caused by other burning materials. [32]

Wibbenmeyer, et al., conducted a review of 194 flame burns from an Iowa burn unit that had been incurred when the victims were burning waste or brush from 1989 to 2000. Such burns accounted for one-fifth of the flame burns treated in an Iowa burn unit during that time. These burns were eight times as common among males as females. Fifty-three percent of these injuries were incurred while burning brush and 29 percent occurred when trash was being burned. Accelerants were involved in 80 percent of these injuries; gasoline was involved in 90 percent of the accelerant injuries. Ninety percent of the brush burning burns involved accelerants, compared to 65 percent of the burns from trash burning. Older victims were less likely to use accelerants than younger ones. Six percent of the trash burns resulted from materials in the rubbish that exploded. Six percent of the victims of brush or trash burning fires died. Elderly victims had the worst outcomes after treatment. Prevention strategies include efforts to reduce the use of accelerants with outdoor burning and establishment of other means of disposing of brush and trash besides burning. [33]

Wildland Fires

Wildland fires are a part of the rural fire problem, although they are not limited to rural areas. Bailey and Montague describe three sets of circumstances where structure and wildland fires intersect. A mixed interface has scattered structures and/or isolated homes in an undeveloped rural area. The risk to individual homes in these areas is high. The occluded interface consists of wildlands, such as a park or conservation land, in an urban area. In a classic interface, a number of homes, such as a subdivision, abut wildlands along a wide front. The classic interface poses a risk of a higher loss of life. [34]

Stephen Badger described four multiple death wildland, wildland-related, or agricultural fires that occurred in 2003. An October 2003 California wildland-urban interface fire of undetermined cause spread across 208,000 acres and killed 13 civilians and one firefighter. A separate, intentionally set California wildland-urban interface fire the same month spread through 91,000 acres and claimed six lives. In March of 2003, six people were fatally injured when a legally permitted agricultural fire was set on a sugar cane field. Warnings had been given, but the six were either asleep or hiding. In August of 2003, eight wildland firefighters returning to Oregon from an Idaho fire died after their van collided head-on with a tractor trailer truck and burst into flames. [35]

The National Association of State Foresters Core Team reviewed the needs and roles of local rural and volunteer fire departments in wildland-urban interface fires. Volunteer firefighters are often the first to respond to wildland or wildland-urban interface fires. The wildland-urban interface requires firefighting equipment, training, and skill in structural and wildland firefighting. Evacuations may be required; communication and interagency coordination are critical. Plans and policies on incorporating local firefighters in multi-jurisdictional responses inside and outside their immediate areas are required. [36[

Rural Fire and Burn Injuries Compared to Other Injuries

Two articles provide a context for fire and burn injuries compared to other types of injuries. Nordstrom, et al., reported on interviews about accidents or injuries within the past year conducted with more than 1,600 residents of an all-rural Iowa county. Twenty-three percent of the people reported an injury in the past year. Burns were the diagnosis in 3.4 percent of the injuries. Fires or burns caused 2.1 percent of the injuries. The percent reporting injuries decreased with age. Women who scored high on the depression scale were more likely than other women to have suffered an injury in the past year. Men who were dependent on or abused alcohol were more likely to have had an injury than other men. The authors also noted that in their study, 48.5 percent of the injuries occurred while the individual was working, compared to 26.3 percent found nationally by the U.S. National Health Interview Survey. [37]

Baker, Whitfield, and O'Neill mapped 1979 to 1981 injury death rates by county for all unintentional injury deaths, firearm homicides, suicides, house fires, drowning, and drowning deaths of children under 5 years old. They found that the West and the South had the highest death rate from unintentional injuries, with the highest rates in rural counties. The rate was much lower in the Northeast. The South, with the exception of

Florida, had very high rates of deaths from house fires with especially severe problems in the Mississippi River Valley and Atlantic coastal low-income counties. House fire death rates were generally low in the Western part of the country. The authors note that noncentral heat is more common in low-income areas of the South, more older wooden buildings are found in the South than in most parts of the country, and distances are greater for rural firefighters. [38]

Programs In or Targeting Rural Areas

Most of the programs or issues discussed in the remainder of this review do not address fire specifically. However, they can suggest avenues to try or partnerships to explore.

Bull and Bane discussed issues of mental health program development in rural settings. Geographic isolation, economic deprivation, human service infrastructure, and economies of scale must be addressed. The distance that must be traveled is a part of the isolation. Public transportation is close to nonexistent. Terrain and weather can also make driving difficult. Costs for long-distance calls, fuel, and travel time add up quickly.

Economic deprivation is exacerbated by the tendency of many rural areas to rely heavily on one industry, activity, or service for local livelihoods. Economic shifts or plant closures can be devastating to residents' income and local tax revenues. Some Federal and State programs mistakenly assume that services can be provided at less cost in rural areas and do not fund adequately. Because rural incomes are lower, and fewer foundations are rural, fewer charitable resources exist for programs or for the matching funds necessary to qualify for some grants.

The human service infrastructure has experienced consolidations and closings. There is a shortage of technical equipment and skilled personnel. Rural youth often move away, and more women are working, reducing the volunteer pool that might partially alleviate the lack of paid workers.

Economies of scale are difficult because the numbers of people and suppliers are simply not there. In some cases, there is a sole supplier. Competitive bidding may not be possible.

The independence associated with rural life, particularly among the elderly, often results in a resistance to using or accepting services or assistance. Rural residents may be suspicious of newcomers or outsiders and fear that State or Federal services could usurp local control. Privacy is also an issue in rural communities.

Seven points are made about transferring urban programs to rural areas:

1. Expectations may need to be scaled, back, particularly if success is defined as number of people served.

2. It often is necessary to scale services to offer only the highest priority (as defined by the community), rather than offering the full range.

3. Program duplication should be avoided, and offerings coordinated so that each agency offers programs it can do best.

4. Rules and regulations should be handled with some flexibility, as bookkeepers and accountants tend to be in short supply. Budget waivers should be sought when expenses will be higher than expected for items like long-distance calls and mileage.

5. Do not expect economies of scale or more than one provider bidding.

6. Create partnerships or reciprocal agreements so that the jurisdictional or administrative boundaries do not interfere with services.

7. Plan for challenges in recruiting and keeping qualified personnel. Hire people with multiple competencies rather narrow specialists.

Many services are delivered without benefit of formal office space. These services may be delivered from stores, churches, restaurants, or vehicles. Gatekeepers, including mail carriers, beauticians, and neighbors, can be used for referrals. The cooperative extension network is recommended as a vehicle for educational programs. Aging, nutrition, and hospital programs can support and publicize a new endeavor. Programs that operate in isolation are less likely to be successful. [39]

Bane and Bull also discussed mental health service delivery for rural elders. They found that, in many programs, one individual's enthusiasm, work, and commitment were critical in organizing and persuading others to establish a program. A credible "natural leader" from the community knows how to present the concept in a way that would be acceptable locally and could motivate other groups to participate.

Services had to feel comfortable to clients. Impressions of comfort can be based on the program's appearance, location, time, and expense. Some programs provided services through nontraditional but trusted partners such as grain dealers, banks, and utility providers. Flexibility to use portions of the services when they choose was important. Rural elderly women tended to be cautious about accepting formal services and were hesitant to accept when they felt that they could not reciprocate.

Two direct service models were described: gatekeepers and peer counseling. Gatekeepers routinely have contact with people who themselves would not seek services. These gatekeepers make referrals to appropriate agencies. Elders were more open to peers than professionals whom they feared might threaten independence. Peers often have a more realistic understanding of the client's situation.

The authors noted the importance of rural-specific material. Crisis intervention materials that advise calling 9-1-1 are not appropriate in areas that lack that service. The sheriff may be the emergency contact. [40]

Programs and Potential Partners in Rural Areas

USDA loans and grants

The USDA Home Repair Loan and Grant Program (Section 504) offers low-interest loans of up to $20,000 and, for people who are at least 62 and unable to repay a loan, grants of up to $7,500 for repairs or the removal of health or safety hazards. Loans also may be used for improvements or modernization. The USDA, through its community facilities program, also coordinates the Rural Emergency Responders Initiative. This program offers financial assistance for equipment, vehicles, and/or buildings, for fire, police, heath care, and other activities. Priority for grants is given to low-income communities and communities with populations under 5,000. Health care, public safety, and community or public service projects also receive priority. Loan programs also are available for rural areas and small towns with populations up to 20,000. [41]

Micro Rural Fire Department program in Alaska

Project Code Red, or the Micro Rural Fire Department, was developed to address the fire problem and lack of firefighting equipment in the small rural areas of Alaska. This project uses new and existing technologies and State-certified training to cope with the extreme winter temperatures, lack of hydrants and, in many cases, lack of roads. The program equips five firefighters with fully-supplied trailers that can be transported by all terrain vehicles, snow machines, pick-up trucks, or by hand, even on boardwalks and trails. The trailer is shipped in a heated and insulated container that doubles as a firehouse. Six hundred gallons of environmentally safe foam for firefighting are provided. The unit can be recharged in less than 5 minutes for less than $50. State-certified firefighter training, based on an adapted version of NFPA Firefighter I, also is included. The training is intended for fire departments without protective gear that have only a limited water supply and may have only portable extinguishers and pumps. The total cost is about 70 percent less than the cost of a new fire engine. [42]

African-American churches in rural communities

Stephanie Boddie discusses the role of African-American churches in rural communities, and particularly in the largely African-American town of Boley, Oklahoma. Historically, many congregations have provided some type of social services, including visits to the sick and other health programs, international relief, educational or cultural activities, environmental issues, and food distribution. Boddie references Chaves and Higgins who found that African-American congregations were more likely than white congregations to be involved in civil rights and providing basic needs to the immediate community. Rural churches are described as similar to an extended family. These rural churches are important forces in leadership development and community organizing. They also maintain strong interactions with the private sector. In both rural and urban African-American communities, the churches are among the chief sources of influence and support.

Four hundred people live in the town of Boley proper and over 900 live in the surrounding area. Most are African-Americans. The average church in the community has 22 active members with roughly one-third of the members 65 or older, and 39 percent ages 25 to 64. Three out of four congregants have annual household incomes of under $25,000; one-fourth of the annual incomes are between $25,000 and $50,000 per year.

The churches are described as the "most prominent and well-maintained structures in town." (p. 325) The churches often provide members, leadership, and resources to social programs independent of government policies. These churches sponsor a wide variety of programs such as literacy, drug and alcohol prevention, and youth outreach. They also support many other programs such as the library, Scouting, and youth recreation. Fourteen of Boley's congregations are members of a ministerial alliance; two of the three remaining assist when called upon. The alliance "has been most effective in maintaining and institutionalizing the programs initiated by the various participating churches, including a community choir, a senior center, funeral ushers, a crime watch group, financial assistance, historic preservation, hosting of holiday celebrations and civic group meetings, sponsorship of the Sunday School Institute, vocational training, a volunteer fire department, and a literacy program that culminates in a GED." (p. 328) The 15 bi-vocational ministers donated time and brought skills and networks from their other careers to social service provision. [13]

Community organizing efforts at a rural manufacturing plant

Fires, Ripley, Figueiredo, and Thompson describe community organizing efforts in smoking cessation and dietary improvements at a rural manufacturing facility in Mecklenburg County, Virginia. The county was 80 percent rural. The population was described as "underserved in terms of medical care, patient education, and cancer education." (p. 414) Four manufacturers were considered for the program. The site that was chosen was operating around the clock and had a reputation as a well-run company with a stable work force. Most employees at this site were African-American and had a high school education.

The project began with the recruitment of a Health Advisory Board (HAB) from the employees. A dozen volunteered at the outset to be on the HAB. The HAB was assigned to "identify and develop activities and health promotion ideas to address cancer prevention and changing diet and smoking among their co-workers." (p. 415) Project staff attended the monthly meetings at the site.

With support from the project, the HAB conducted one activity per month over nine months, usually around shift changes. Completion rates were high for short surveys before and after the program.

Some information (generally printed) was given to all employees. Thirty-five percent attended a dietary fat screening. Before the program began, 40 percent of the survey respondents were smokers, compared to 35 percent at the end. Only four percent attended a nutrition seminar that was offered twice. A "stop smoking" contest attracted 13, or 18 percent, of the known workplace smokers. All 13 were able to quit for 24 hours; five were still not smoking after 2 months. In addition, smokers at the plant tend to be somewhat readier to quit smoking at the end of the program than they were at the outset. Daily consumption of fruit and vegetables increased and employees became more self-confident in reducing fat consumption.

Costs for the total program came to about $1,925. This does not include donations of pamphlets, posters, and prizes for the "stop smoking" contest or the time spent by employees on this project.

The authors conclude that community organizing strategies at the workplace may be appropriate to reach minority rural residents. "A low-intensity community organizing approach with minimal intervention resources can reach employees in such work sites and produce small behavioral and attitudinal changes." [44]

Health promotion programs by "predominantly rural" North Carolina hospitals

Christine Dorresteyn-Stevens described the health promotion programs offered by North Carolina hospitals with fewer than 100 beds serving predominantly rural populations. Twenty-nine of the 45 such hospitals in the State responded to the survey.

Ninety-three percent of the rural hospitals offered at least one health promotion program. At least half of the responding rural hospitals offered first aid and cardiopulmonary resuscitation, AIDS education, nutrition, prenatal education, and breast self-examination. Other common programs included smoking cessation, weight control, and stress management. Hospital employees were the most common target audience with non-patients and members of the community ranking second. Some programs were provided for in- and out-patients. The target audience varied by program. A variety of financing methods were used. Nurses were the primary coordinators of this programming at 85 percent of the small hospitals.

Health promotion programming was sometimes considered part of public relations. Programs targeting hospital employees may be developed for certification, to reduce absenteeism, or to increase productivity. Most program sessions were conducted by regular staff. Having programs generally coordinated through nursing or in-service departments increases the likelihood that these programs will be for hospital staff and may make the development of programs for the community less likely.

Dorresteyn-Stevens noted that hospitals could serve as links between health care, businesses, and community agencies. These institutions could be key players in establishing a group to coordinate health promotion activities in the area. The inclusion of community residents can increase participation and better identify the priorities of the particular community. [45]

Twelfth Grade Students as Rural Health Educators

Whitener referenced Pye, O'Loughlin, Dodson, and Pye's work describing a rural community health education project, "HeartSmart," in which 12th graders were trained in cardiovascular disease prevention by a local hospital, and then developed a program and taught sixth-grade students about cardiovascular health. The posttest showed an increase in knowledge. The new partnerships were also considered beneficial. [7]

Annotated Bibliography

1. Rural Assistance Center. "What is Rural? Frequently Asked Questions" accessed online at *http://www.raconline.org/info_guides/ruraldef/ruraldeffaq. php#definition* on May 31, 2005.

The Rural Assistance Center, developed by the U.S. Department of Health and Human Services' Rural Initiative, discusses how "rural" is defined. Some definitions are context-driven. Areas considered rural in densely populated States, such as Massachusetts, can have little resemblance to areas described as rural in less densely populated States, such as Montana.

The U.S. Census Bureau, the Office of Management and Budget, and the Economic Research Service of the USDA all have definitions of rural that are widely used.

The U.S. Census Bureau considers areas other than urbanized areas or urban clusters to be rural. An urbanized area has a nucleus (may or not be a unique city) with at least 50,000 residents. Such an area also has a core of at least one contiguous block group of less than 2 square miles with 1,000 people per square mile. Urban clusters have similar cores, but they have populations of 2,500 to 49,999.

Metropolitan statistical areas, as defined by the Office of Management and Budget, include "central or core counties with one or more urbanized areas, and outlying counties that are economically tied to the core counties as measured by work commuting." Micropolitan statistical areas include a) nonmetropolitan counties with at least one urban cluster of 10,000 or more residents, and b) noncore counties that lack these urban clusters. Both types of nonmetropolitan counties are often included in studies of rural conditions.

Several classifications have been devised by the Economic Research Service (USDA). Census tract classifications combined with Census definitions of urbanized area and place and commuting information are used to define Rural-Urban Commuting Areas (RUCAs).

2. Weiner, Robert J., and Joseph N. Belden. "The Context of Affordable Housing in Rural America." Ch. in *Housing in Rural America*, Joseph N. Belden and Robert J. Wiener, eds., California: Sage Publications, 1999, pp. 3-12.

In this discussion, rural housing is found in small towns and open country. Belden and Weiner cite Wilson and Carr, other authors in the same text, who list "remoteness, low population density, and economic dependence on a single industry" as key features. Rural housing includes housing in areas specifically classified as rural and "other urban." The American Housing Survey (AHS) used the 1980 U.S. Census definitions for urbanized areas (incorporated places and densely settled surrounding areas [at least 1.6 people per acre] with a combined population of at least 50,000). "Other urban areas" are areas with populations of at least 2,500 that are not inside the urbanized areas. Rural housing included that which was "not classified as urban." Counties that fall outside of metropolitan statistical areas are considered nonmetropolitan. Metropolitan areas include counties having central cities with populations of at least 50,000 and surrounding metropolitan-type counties. Metropolitan areas may have portions that meet the rural definition of communities with fewer than 2,500, and nonmetropolitan areas may have rural and urban sections, the later with populations of 2,500 or more. In 1995, 74 percent of nonmetropolitan and 82 percent of rural housing was owner occupied, as compared to 59 percent of urban and 49 percent of the housing found in central cities. Roughly three-quarters of the housing was single-family detached.

3. Ahrens, Marty. *The U.S. Fire Problem Overview Report: Leading Causes and Other Patterns and Trends.* Quincy: NFPA, June 2003, pp. 27-31.

During the 5-year period 1997 to 2001 (excluding the events of September 11, 2001), the average death rate of 30.9 deaths per million population for rural areas (populations under 2,500) was at least twice that seen for all other population intervals except for the 2,500 to 5,000 interval, which averaged 18.4 deaths per million population. Rural communities averaged 12.0 reported fires per 1,000 population, twice that of all population intervals except for 2,500 to 5,000 (8.3), and 5,000 to 10,000 (6.9). Per capita death rates are at the lowest point in communities with populations of 50,000 to 99,999. Fires per capita hit their nadir in communities of 25,000 to 49,999 in population. The rates go up again slightly for larger cities, although these city rates remain far below the rural rates. Across the board, there is less difference in the rate of fire deaths per 1,000 fires than in population rates.

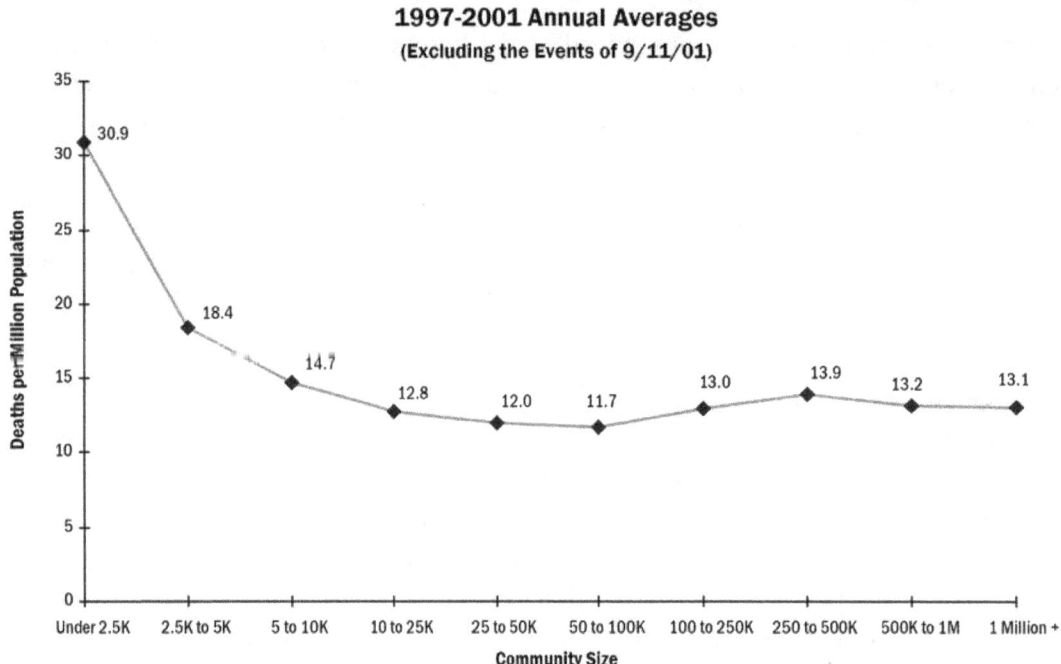

Civilian Fire Deaths per Million Population by Size of Community:
1997-2001 Annual Averages
(Excluding the Events of 9/11/01)

Source: *Fire Loss in the United States,* by Michael J. Karter, Jr.

Death rates by community size tend to rise and (more often) fall together, showing considerable year to year variation in the size of the problem, but much less variation in the rankings of different sized communities by death rate. Sharp upward rises were seen in the 1992 and 1996 fire death rates for rural, but these primarily reflect the much larger statistical uncertainty associated with 1-year estimates.

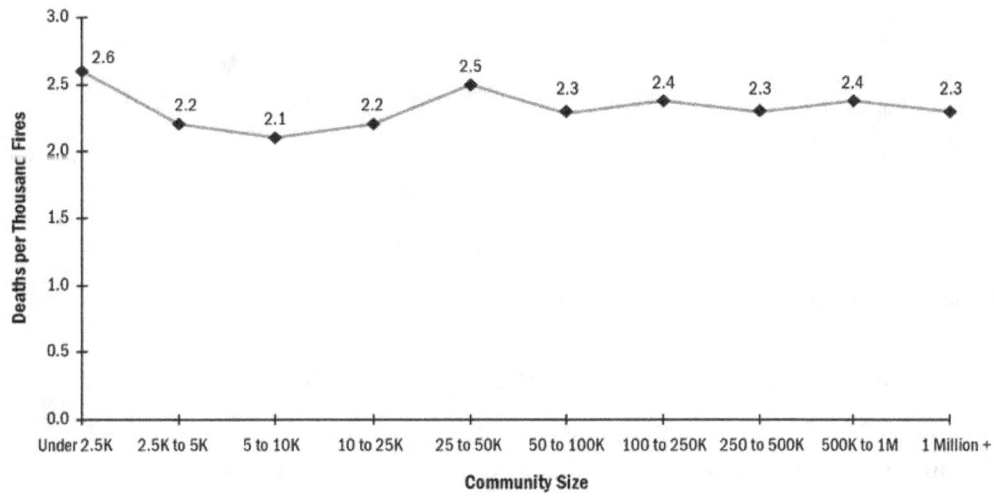

**Civilian Fire Deaths per Thousand Fires by Size of Community:
1997-2001 Annual Averages**
(Excluding the events of 9/11/01)

Source: *U.S. Fire Experience by Region,* by Michael J. Karter, Jr.

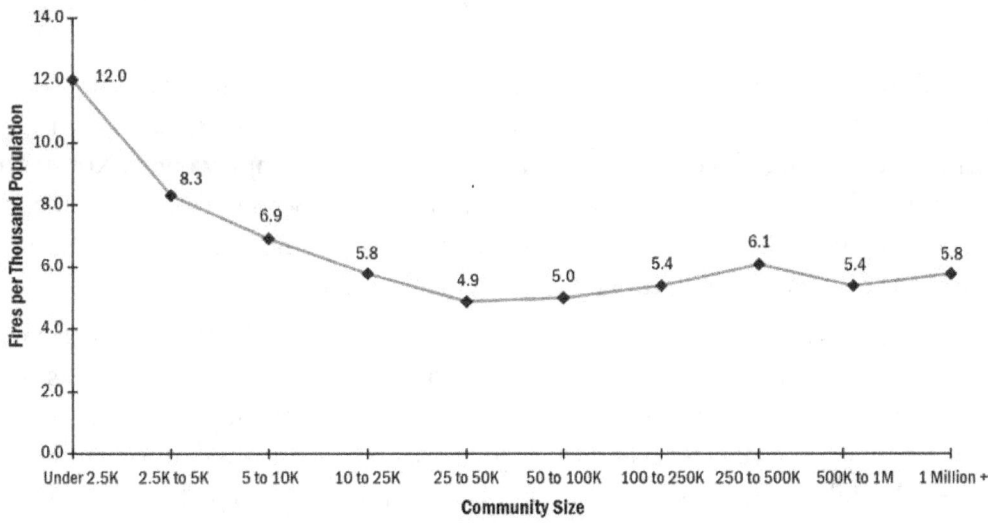

**Reported Fires per Thousand Population by Size of Community:
1997-2001 Annual Averages**

Source: *Fire Loss in the United States,* by Michael J. Karter, Jr.

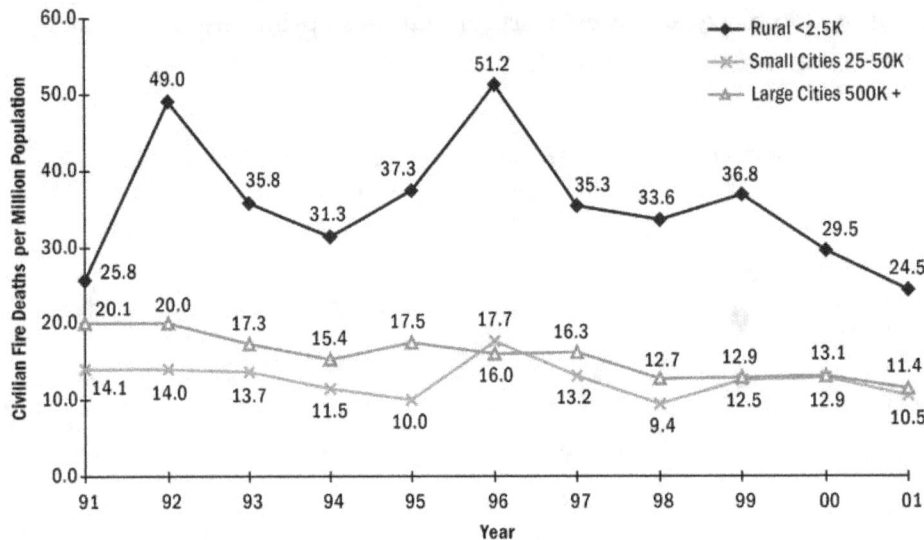

Trends in Civilian Fire Death Rate per Million Population by Community Size: 1991-2001

(Excluding the Events of 9/11/01)

Source: *Fire Loss in the United States,* by Michael J. Karter, Jr.

4. Northeast Midwest Institute. "2000 Rural Population as a percent of State Total By State," 2002, accessed online at *http://www.nemw.org/poprural.htm* on May 18, 2005.

According to U.S. Census figures, 59 million, or 21.0 percent, of the U.S. population is rural. Three million, or 1.1 percent, of the U.S. population lives on farms. Fifty-six million, or 19.9 percent, live in nonfarm rural areas. The table on the Web site shows the total population, rural population, rural population percent and percent ranking, and farm and nonfarm rural populations and percents.

5. Hall, John R., Jr. *U.S. Fire Death Rates by State.* Quincy: National Fire Protection Association, Oct. 2004.

This analysis examines State fire death rates and trends in connection with other factors, such as poverty, smokers, and lack of education. Death certificate data from the National Center for Health certificates with external cause of injury codes X00-09 for 1999 to 2001 and E890-899 for earlier years were included. Some fire deaths, including those caused by vehicle fires and fire-related homicides and suicides are excluded. Deaths from the events of September 11, 2001, also are excluded.

Almost all States have shown drops in fire deaths and fire death rates over the past 2 decades. The Southeastern States of the old Confederacy (excluding Florida), plus Alaska, have consistently had fire death rates above the national average. Border States such as Missouri and Oklahoma also tend to have high rates. States with small populations can have unusually high death rates in some years due to an increase of just a few deaths.

The percentage of State residents with incomes below the poverty line, of adults with less than 12 years of school, and of adult smokers were correlated with high death rates. These factors also are correlated with each other. By themselves, education could account for 29 percent of the difference in fire death rates, smoking prevalence 22 percent and differences 16 percent. According to current data, poverty accounts for a smaller share of the difference in the fire death rates than in the past. Several States with large populations of Native Americans have risen in the poverty ranks. It is possible that Native American fire deaths are not included as completely in the death certificate database as Native American poverty is in the poverty statistics.

Larger percentages of African-American populations also are correlated with higher fire death rates. However, education and poverty have stronger predictive power.

Where the housing has been maintained, age of housing is a poor predictor of fire death rates. Older housing is at higher risk for electrical fires.

Colder climates are not associated with higher fire death rates. The Southeast has the highest death rates from home heating fires because of the lack of central heat and reliance on portable and area heaters or makeshift arrangements.

6. Dolbeare, Cushing N. "Conditions and Trends in Rural Housing." in *Housing in Rural America,* Joseph N. Belden and Robert J. Wiener, eds., California: Sage Publications, 1999, pp. 13-26.

Dolbeare considered small towns and open country rural. In terms of census definitions, this would include rural and "other urban" (at least 2,500 residents outside of urban areas). In 1995, according to these definitions, 37.2 million (38 percent) of the Nation's 97.7 million housing units were rural. Forty-four percent of the rural units were in metropolitan areas and the remainder were outside metropolitan areas. Metropolitan areas, outside of New England, are defined in terms of whole counties.

Seventy-six percent of rural householders owned their own home, compared to 58 percent of urban dwellers. Three-quarters of rural housing units were single-family homes. Roughly one-eighth of the rural householders live in manufactured housing. Only 11 percent live in properties with more than one unit. In rural areas, householders tended to be older, poorer, and more likely to be married and white than their urban counterparts. Although rural and urban poverty rates were similar, 22 percent of the rural households had incomes between 100 percent and 200 percent of the poverty level. However, housing in rural areas tends to be larger and less expensive than urban housing. Rural renters face a higher housing cost burden than homeowners.

Five percent of rural African-American households, 3 percent of rural Hispanic households, and 2 percent of rural white households lived in severely inadequate housing. Seventeen percent of rural African-American households, 10 percent of rural Hispanic households, and 4 percent of white households lived in housing that was considered moderately inadequate.

Twenty-seven percent of all rural householders were at least 65 years old. Twenty-one percent of these older rural householders had incomes below the poverty line, and 31 percent had incomes between 100 percent and 200 percent of the poverty level.

Single parents accounted for one in ten rural householders. Forty-eight percent of the single parents were renters, and 43 percent had one or more children under six years old. Thirty-four percent of the rural single parent households had incomes below the poverty line, and 29 percent had incomes between 100 percent and 200 percent of the poverty level.

7. Whitener, Lynn. "Families and Family Life in Rural Areas." *The Journal of Rural Health* 11, 3 (Sept. 1995): 217-223.

This research review contains an overview of work done by others authors describing demographic data, resource availability and effect, and health care issues. The author notes that "Poverty, isolation and limited options are the reality of rural life for children." (p. 217) A 1994 study by First, Rife, and Toomey of nonurban and rural Ohio homeless found a larger share of young women with children who became homeless due to family conflict than had been seen in national studies. They also found that homeless, single-parent rural families were more likely to rely on relatives for resources than comparable families in urban areas who had greater access to social services.

Atkinson (1994) found that rural families were more likely to use relatives for child care, that rural mothers have less education than urban mothers, have more children, and that rural mothers are more likely to be employed in clerical work than are urban mothers. Rural poverty has increased with changes to the structure of the family. Rural fathers in two-parent families have lower incomes than urban fathers.

Stoneman, Flor, McCrary, Hastings, and Conyers (1994), studied 90 African-American children between 9 and 12 years old. The children's parents were married, and the families lived in the rural South. More conflict, more depression, and less support were found in families with fewer financial resources. They also found that conflict and "loss of family optimism" (p. 217) were associated with a lack of youth self-control or self-discipline and diminishing achievement.

Pye, O'Loughlin, Dodson, and Pye's 1993 work described a rural community health education project, "HeartSmart," in which 12th graders were trained in cardiovascular disease prevention by a local hospital, and then developed a program and taught sixth grade students about cardiovascular health. The posttest showed an increase in knowledge. The new partnerships were also considered beneficial.

Houston, Meade, and Mainous (1992) found that rural Kentucky physicians were less likely to offer immunizations than were urban doctors. Both urban and rural doctors reported that cost was the problem and would refer patients to county health agencies.

Terwilliger (1994) provided an overview of circumstances leading to the earliest health clinics in schools. Pediatric, school nurse, and school health groups supported the idea while conservative groups and individuals sometimes opposed them because of concerns about sexuality or parental rights. Reimbursement was a major issue.

Lehmann, Barr, and Kelly (1994) analyzed adolescent patients seen by a rural hospital emergency room. Many lacked primary care physicians. Half had private insurance. Those who had such insurance were more likely to have a primary doctor.

Acheson's 1994 study of Old Order Amish found a rural population that did not use contraception, had many children close in age, and valued women's roles. Tobacco and alcohol use is prohibited. Their strong social network also provides financial support.

Wheeler and Lewis (1993) found that rural families caring for medically fragile children at home could find their situations further complicated by isolation, power outages, limited emergency services, and the need to use well water and septic systems instead of municipally provided services.

Whitener notes that the issue of self-reliance is seen repeatedly in these papers. Greater interdependence is seen in the absence of the social supports found in more populous areas. Rural advocates are trying to bring more of these social resources to rural communities.

8. Belden, Joseph N. "Housing for the Rural Elderly." Ch. in *Housing in Rural America,* Joseph N. Belden and Robert J. Wiener, eds., California: Sage Publications, 1999, pp. 91-97.

In 1995, 13.9 percent of the residents of nonmetropolitan areas were at least 65, compared to 11.9 percent of the population overall. Forty-six percent of the nonmetropolitan African-Americans who were at least 65 lived in poverty. Thirty-three percent of the nonmetropolitan Hispanics lived in poverty. Nonmetropolitan areas had larger shares of older adults with incomes under 200 percent of the poverty level.

Compared to elders elsewhere in the country, homes of older adults in nonmetropolitan areas were more likely to be owner-occupied and to have problems. The nonmetropolitan older adults were more likely to live in manufactured housing than their metropolitan counterparts and to have fewer economic resources. In 1995, more than one million nonmetropolitan housing units occupied by older adults lacked adequate heating equipment. Almost half a million (468,000) nonmetropolitan elder-occupied units had severe or moderate problems with heating, plumbing, electrical systems, maintenance, kitchens, and/or hallways.

Rental housing is less available in nonmetropolitan and rural areas. While many prefer to stay in their own homes, having landlords or property management handle maintenance and repair can make sense to people who no longer want these responsibilities.

The Rural Housing Services offers grants and low-interest loans available to very low-income rural residents. These funds can be used to install indoor plumbing, upgrade heating, or make necessary repairs.

9. Kumar, Vinod, Miguel Acanfora, Catherine Hagan Hennessey, and Alex Kalache. "Health Status of the Rural Elderly." *The Journal of Rural Health* 17, 4 (Fall 2001): 328-331.

This study provides a brief international overview of health issues facing rural elderly. They reference work from Hennessy, Moriarty, Zack, Scherr, and Brackbill (1994) about health risk behavior in rural and urban Americans of at least 60 years of age. Using

1993-1998 data extracted from the CDC's Behavioral Risk Factor Surveillance System (BRFSS), they found that 18.9 percent of rural Americans 60 and over were obese, 43.1 percent were physically sedentary and 14.6 percent were current smokers. For urban Americans of the same age, 17.6 percent were obese, 37.4 percent were physically inactive, and 13.1 percent were current smokers.

10. McCray, Jacquelyn W. "Affordable Housing in the Rural South." Ch. in *Housing in Rural America,* Joseph N. Belden and Robert J. Wiener, eds., California: Sage Publications, 1999, pp. 29-48.

Poor housing quality was identified as a problem in the rural South in the 1940s and has remained a concern. Three-quarters of the substandard housing units in the 1980s were in the South. In 1995, the 9 percent of the Nation's housing units in the non-metropolitan South accounted for 21 percent of the Nation's supply of occupied units with moderate physical problems, 11 percent with severe problems, 26 percent of the country's manufactured housing, and 12 percent of the households with income below the poverty level. Thirty percent of the people in the 16 Southern and border States lived in rural areas. According to the 1990 Census, 22 percent of the rural Southern population had incomes below the poverty line.

In 1990, 2.5 percent of rural Southern housing units did not have complete plumbing. Even larger shares of homes in rural Kentucky, (4.8 percent), Virginia, 4.4 percent, and West Virginia (3.6 percent) lacked these facilities. About 3.9 percent of rural units in the South were overcrowded. Texas (6.4 percent, Mississippi, (6.1 percent) and Louisiana had the largest percentages of overcrowded rural housing units. Mississippi, Texas, and Kentucky had the largest shares of substandard rural housing, with Mississippi, Texas, and Louisiana having the largest share of total substandard housing. Housing problems tend to be greater in the 214 counties of the Lower Mississippi Delta (LMD). The LMD includes counties in Mississippi, Arkansas, Louisiana, Tennessee, Kentucky, Missouri, and Illinois. Larger shares of owner-occupied housing in the LMD had values under $15,000 as compared to other counties in the States. In the LMD region, poor and minority neighborhoods had infrastructure that was inferior to that found in other communities.

11. Wimberley, Ronald C., and Libby V. Morris. *The Southern Black Belt.* Lexington: TVA Rural Studies, 1997.

The Southern Black Belt is a crescent-shaped belt of counties in Virginia, North and South Carolina, Georgia, Florida, Alabama, Mississippi, Tennessee, Arkansas, Louisiana, and Texas that has more than the average percentage of African-American residents.

Although agriculture is an important part of the region's economy, the number of Southern farms has declined. Very few farms are operated by African-Americans. Forty-four percent of the rural population (45 percent of nonmetropolitan population) in the U.S. live in the South. Seventy-nine percent of the U.S. nonmetropolitan African-

American population lives in the Black Belt. Ninety-one percent of all nonmetropolitan or rural African-Americans live in the South.

The 11 States in the Black Belt are over-represented in lists of highest unemployment, poverty, infant mortality, and hunger. The authors note that "The 11 Black Belt States contain 35 percent of the Nation's poor, 43 percent of the nonmetropolitan poor, 51 percent of the African-American poor, and 90 percent of the nonmetropolitan African-American poor. Within these 11 States, the 623 Black belt counties claim 23 percent of all U.S. poverty, 28 percent of the nonmetropolitan poverty, 47 percent of the African-American poverty, and 84 percent of the nonmetropolitan African-American poverty." (pp. 8-10) White poverty in this region is less prevalent and scattered along the Louisiana Delta and lower portion of the Black Belt. Areas of the country with high levels of white poverty (the Appalachians, the Missouri-Arkansas-Oklahoma area, Southwest Central, and Upper West) tend not to have high levels of African-American poverty.

Forty percent of U.S. residents who lack high school diplomas live in the South, with 21 percent in the Black Belt. The majority of counties in the Black Belt and the rest of Florida are in the top quartile in percent of African-Americans without high school diplomas. Counties in the top quartile of whites without diplomas are more common in the upper South from the Appalachian Mountains to Oklahoma. A strong correlation exists between the lack of education and poverty.

12. Harris, Rosalind P., and Julie N. Zimmerman. "Children and Poverty in the Rural South." *SRDC Policy Series* no. 2 (Nov. 2003), accessed online at *http://srdc. msState.edu/publications/srdcpolicy/harris_zimmerman.pdf* on May 31, 2005.

In 2001, 16.3 percent of U.S. children lived in poverty. The poverty rate was higher (20.3 percent) in nonmetropolitan areas than in metropolitan areas (15.4 percent). In the South, 18.9 percent of all children lived in poverty, with 17.3 percent of the metropolitan children and 24.9 percent of the nonmetropolitan children living under these conditions. The Midwest had the lowest child poverty rate for all three conditions.

Child poverty rates also very by race and ethnicity. Nationally, 40.5 percent of African-American children, 17.0 percent of white children, and 32.3 percent of the Hispanic children living in nonmetropolitan areas of the country were living in poverty. In metropolitan areas, 28.7 percent of the African-American children, 12.6 percent of the white children, and 27.5 percent of the Hispanic children lived in poverty. Interestingly, the South had slightly lower percentage of African-American (39.8 percent) and Hispanic (31.9 percent) nonmetropolitan child poverty, while the 19.8 percent white nonmetropolitan child poverty was slightly higher.

One part of the country known as the Black Belt extends through East Tennessee, the Carolinas, Virginia, Alabama, Georgia, Mississippi, Arkansas, Louisiana, and East Texas. Forty-six percent of the African-American population of the U.S. and 83 percent of the nonmetropolitan African-American population live in this region. Thirty-four percent of the poor in the U.S. live in this region, including 43 percent of the country's rural poor.

In 1990, the Black Belt was home to 90 percent of the nonmetropolitan rural African-American poor.

The rural South evolved from an economy with plantations, coal mining, and timber industries to other manufacturing. However, manufacturing jobs have shifted in recent decades to more urban areas or to other countries, and the economy has shifted to services. Investment in education and training has been weak in the rural South, leaving a work force with fewer skills. Working poverty is common with many part-time and seasonal positions. Parents of poor children in the rural South are younger and less educated than parents of poor children elsewhere. Single-parent families face a greater burden. Fifty percent of rural Southern children in mother-only families lived in poverty compared to 16 percent in two-parent families. Family income influences housing quality, neighborhood, and educational and social opportunities. A sense of hopelessness can result from the struggles to survive, and children can develop limited aspirations. Welfare reform has led to many families leaving welfare rolls, but remaining in poverty. Resources are often lacking for child care, transportation, and prevention programs in rural areas. Racial and sexual discrimination has been a factor in excluding minorities and women from educational and employment opportunities that would enable an adequate, stable income. Families lack the resources to break the cycle.

13. Mather, Mark. "Housing and Commuting Patterns in Appalachia." *Demographic and Socioeconomic Change in Appalachia.* Washington: Population Reference Bureau, January 2004, accessed online at *http://www.arc.gov/images/ reports/housing/Housing_commuting.pdf* on May 31, 2005.

This study compared census and other data for metropolitan and nonmetropolitan areas in Appalachia and elsewhere in the U.S., and within Appalachia, and examines data for distressed, transitional, competitive, and attainment counties. In 2000, 75.5 percent of the nonmetropolitan Appalachian housing units were owner-occupied, compared to 72.0 percent of the metropolitan Appalachian units. Home ownership was highest (76.9 percent owner-occupied) in distressed counties, and lowest in the attainment counties (69.2 percent). Manufactured homes accounted for 19.5 percent of the housing units in nonmetropolitan Appalachia, 24.7 percent of the homes in distressed Appalachian counties and only 4.4 percent of the attainment Appalachian counties. (Attainment counties have income, unemployment, and poverty rates that are equal to or better than the national average.) The lower cost of manufactured housing has contributed substantially to the above-average levels of home ownership in the region. Nationally, 66.2 percent of U.S. homes were owner-occupied in 2000. While only 1.0 percent of the nonmetropolitan Appalachian homes had incomplete plumbing in 2000, 1.6 percent of the homes in distressed counties had incomplete plumbing. The highest percentage of Appalachia homes without complete plumbing was found in Hancock County, Tennessee, with 8 percent lacking these facilities. The largest shares of Appalachian homes with incomplete plumbing were found in Kentucky and West Virginia.

In 2000, 2.4 percent of all U.S. households had no access to telephones, but 4.7 percent of the nonmetropolitan Appalachian households lacked this access, and the percentage lacking telephone access in distressed parts of Appalachia was 6.9 percent. In 23 Appalachian counties, 10 percent or more the households did not have access to telephones in 2000. The data do not indicate whether the problem is lack of funds or lack of available phone service in the community.

14. Krofta, Janet A., Sue R. Cull, and Christine C. Cook. "Affordable Housing in the Rural Midwest." Ch. in *Housing in Rural America,* Joseph N. Belden and Robert J. Wiener, eds., California: Sage Publications, 1999, pp. 61-73.

In recent years, many people have left the rural Midwest, leaving an aging population and an aging housing stock. In the Midwest and the rest of the country, nonmetropolitan households are more likely to be white, older, and married. In 1995, 28 percent of the nonmetropolitan Midwest households were headed by someone at least 65 years of age compared to 21 percent nationally. Nationally, the median year of construction for housing was 1986, in the nonmetropolitan Midwest, it was 1968. Housing costs are lower in the rural Midwest than in the country as a whole. One-family, detached dwelling units accounted for 79 percent of the nonmetropolitan Midwest housing compared to 68 percent of the Nation's housing supply. Manufactured homes accounted for 8 percent of the nonmetropolitan Midwest homes and 6 percent of the country's housing. Twenty-five percent of nonmetropolitan Midwest housing is at least 50 years old, and much of this housing needs updating. Only the Northeast has more older housing, with 33 percent of its housing stock that old.

Slightly more than half of nonmetropolitan Midwest homes use piped gas as their main source of heat, and two-thirds were getting water and sewer services from a water company or public system.

Consistent with other areas of the country, households of older adults, and those headed by women and/or minorities were most likely to have inadequate housing and to be larger shares of their income for housing. Native Americans and migrant farm workers in the region are confronted with especially severe housing conditions. They are also frequent victims of discriminations. Migrant workers face the added challenge of frequent moves.

Farming no longer dominates the rural Midwestern economy. In almost all nonmetropolitan Midwest counties, agriculture now accounts for less than 20 percent of the labor and proprietor income. Manufacturing and mining have also declined, as have businesses that support an agricultural community

15. Slesinger, Doris P., and Cynthia Ofstead. "Economic and Health Needs of Wisconsin Migrant Farm Workers." *The Journal of Rural Health* 9, 2 (Spring 1993): 138-148.

Most of the roughly 2.5 million hired farm workers are seasonal employees who live at home year-round and whose farm labor supplements other income. Migrant workers get most of their annual income from farming, usually travel in families, and may be paid as families. As of 1985, 159,000, or roughly 6 percent, of paid farm workers were migrants. This study compares the characteristics identified by interviews with migrant Wisconsin workers conducted in the summers of 1978 and 1989. The workers tended to work in either the fields or the canneries.

In 1972, 60 percent of the migrant workers in the area were men. Men accounted for 72 percent of the workers in 1989. In 1989, 16 percent of the migrant men and 19 percent of the migrant women were functionally illiterate. In 1989, only 8 percent of the men and 14 percent of women migrant workers 25 years of age or older had completed high school. In the same year, 76 percent of the U.S. population 25 or older had completed high school. Sixty-two percent of the 1989 sample were married, and most married couples had children or other relatives as part of their household. Three generations were present in 11 percent of the households.

While all of the migrant workers in the study lived away from home in July and August, roughly 90 percent were back in their home States in the winter months. Employment increases as planting begins, then peaks in the summer. Almost 60 percent were unemployed in the winter.

The median household income for these workers (average household of 5.2 persons) in 1988 was less than half the Federal poverty level and about a fifth of the national household median income. Income was earned by one household member in one-third of the migrant households in 1988, by two earners in one-quarter, and the others had between three and ten earners. Migrant work was the sole income source for 44 percent percent of the households. Forty percent were paid by the piece or weight rather than in hourly wages.

Only 13 percent of the migrant workers described their heath as excellent, compared to 40 percent of the U.S. population as a whole. About one-third of the workers reported backaches. Migrant workers who spoke only Spanish reported more health problems, including many that were stress-related. However, these workers tended to be older than those who spoke at least some English. More than two-fifths of the workers have never had an eye exam. Only 8 percent of those sampled (including 14 percent of those who didn't speak English) in 1989 reported that language was a problem in obtaining health care, more than half (and 81 percent of those who didn't speak English) needed to have someone interpret when they obtained medical care.

Most of the agricultural migrant workers lived in employer-provide housing. One-third of the housing units did not have indoor plumbing in 1989. In these cases, separate bath houses were provided. During the peak season, workers often worked double shifts 7 days a week. Because migrant workers have many different employers, the process of applying for unemployment insurance is difficult. Many farms do not have enough employees to require contributions to unemployment insurance. Migrant worker who are citizens benefit from more social insurance programs. Undocumented workers are more vulnerable.

16. Peck, Susan. "Many Harvests of Shame: Housing for Farm Workers." Ch. in *Housing in Rural America,* Joseph N. Belden and Robert J. Wiener, eds., California: Sage Publications, 1999, pp. 83-90.

The majority of farm workers in the U.S. are now Latino; many of the workers are undocumented. In the 1980s, farm workers in Delaware, Maryland, and Virginia were usually African-American or Caribbean, but as of 1992, 84 percent were Hispanic. African-American farm workers tend to be single men; the Mexican and Mexican-American workers tend to travel as families. Many of these workers have no reading skills in any language.

In areas with long growing seasons, such as Oregon, farm workers often decide to stay, and the communities become more Hispanic. It is estimated that about 700,000 farm workers are hired annually in California, with 92 percent foreign born. Just 9 percent were not authorized to work in the U.S. In 1994, 47 percent of California were migrants compared to 39 percent in 1983. A growing number of farm workers are from the indigenous peoples in Mexico and Central America. Many of these people do not speak Spanish or English.

A 1993 report found that the median personal income for fieldworkers in California was between $5,000 and $7,500. Only 11 percent received food stamps, 2 percent were getting Aid to Families with Dependent Children, and 3 percent received housing assistance.

Some potential sites for housing have been contaminated with pesticides or are near fuel storage tanks. Necessary mitigation measures would increase costs. Public water and sewer services may not be available or adequate in areas where this housing is needed, and finding suitable sites for septic systems is a challenge. Many communities are opposed to multifamily housing in general, and are concerned about potential resource use if migrant workers are encouraged to settle in the community. Colorado growers have created successful partnerships with local and State government to provide housing that encourages the annual return of reliable workers.

Increased enforcement of health, safety, and housing requirements and the associated increase in penalties have coincided with a decline in farmers providing housing for their workers. In 1968, there were 5,000 labor camps in California that were licensed by the State. In 1994, there were only 1,000. Nonemployer organizations are providing more and more of this housing. Labor contracting is becoming more common and direct hiring less so.

17. Martinez, Zixta Q., Charles Kamasaki, and Surabhi Dabir. "The Border Colonias: A Framework for Change." Ch. in *Housing in Rural America,* Joseph N. Belden and Robert J. Wiener, eds., California: Sage Publications, 1999, pp. 49-60.

A colonia, according to the definition used by the 1990 National Affordable Housing Authority Act is "an identifiable community in Arizona, California, New Mexico or

Texas within 150 miles of the U.S.-Mexico border, lacking decent water and sewage systems and decent housing, and in existence as a colonia before November 28, 1990." (p. 50) Other agencies and jurisdictions use different definitions. In 1995, the Texas Water Development Board estimated that roughly 280,000 people lived in 1,193 colonias in Texas, with 60 percent of this population in the four counties of the Lower Rio Grande Valley, counties that rank among the most impoverished in the country. Estimates of the number of New Mexico colonias and residents vary widely, ranging from 15 to 60 colonias, with 14,600 to 100,000 residents. In 1987, a Congressional Research Service (CSR) study found 25,000 in colonias of San Diego County and 11,500 in Imperial County colonias. Many rural Latino communities strongly resemble colonias but are outside of the border area. The 1987 CSR study found 50 to 55 colonias in Arizona.

Because colonias are physically, and generally legally, isolated, basic infrastructure such as water, sewer, and paving lack economies of scale. Many colonias are not in cities and colonias' residents do not have sufficient income for user fees or many taxes. Colonia residents lack political power. They comprise only a small share of the local population in any voting district. Many Americans believe that most of the residents of colonias are immigrants and are opposed to providing services to immigrants. Colonias do not fit neatly into traditional definitions of rural versus urban, are limited to four States, and are often ineligible for programs. Although advocacy by, and services from, community organizations were common in the colonias during the 1960s and 1970s, funding cuts had a major impact on these activities.

18. Stover, Mary. "The Hidden Homeless." Ch. in *Housing in Rural America,* Joseph N. Belden and Robert J. Wiener, eds., California: Sage Publications, 1999, pp. 75-81.

Definitional challenges make measuring the homeless difficult. A family that doubles up with another is generally not counted as homeless, although their situation may be overcrowded and highly unstable. Many of the factors seen in rural homelessness are consistent with what is seen in urban areas: poverty, lack of affordable housing, unemployment, underemployment, substance abuse, physical or mental health problems, and, for women and children, domestic violence. Domestic violence is believed to be a larger part of the rural homeless problem. A lack of public transportation to services and work reduces options in rural areas.

Compared to the urban homeless, the rural homeless population contains more women and families, fewer men, and tends to be younger. Rural areas have fewer minorities than urban areas, and consequently fewer homeless minorities. However, rural minorities face a higher risk of homelessness than do whites. Native Americans and migrants are at higher risk.

Most of the resources to address issues of homelessness have been spent in urban areas. What resources exist tend to be spread out, and distances pose challenges. Rural communities often do not have hospitals or social service organizations. There is less infrastructure to use.

19. The National Center on Addiction and Substance Abuse at Columbia University. *No Place to Hide: Substance Abuse in Mid-Size Cities and Rural America,* 2000, accessed online at *http://www.casacolumbia.org/supportcasa/item. asp?cID=12&PID=116*

Substance abuse has been associated with large cities, but it is a serious problem in rural areas that frequently lack enforcement and treatment resources found in more populated areas. Rural eighth graders were more likely to have used marijuana (11.6 percent), amphetamines (5.1 percent), cigarettes (26.1 percent) and alcohol (28.1 percent) in the past month than were students in small (marijuana—9.4 percent, amphetamines—3.1 percent, cigarettes—16.0 percent, alcohol - 23.4 percent) and large metropolitan areas (marijuana—8.6 percent, amphetamines—2.5 percent, cigarettes—12.7 percent, alcohol—21.7 percent). Use of cocaine and amphetamines by rural tenth and twelfth graders was higher than use by students in large urban areas. In the seventies and eighties, cocaine use by 12th graders was more common in the large cities. In 1999, rural use by these teens was two percentage points higher than use by 12th graders in large metropolitan areas.

Among adults, marijuana use is less common in rural areas, but no statistically significant differences were seen in use of other illicit drugs. Rural adults were more likely to use tobacco products than were urban adults.

In 1992, Native Americans were found to have much higher rates of illicit drug use, alcohol consumption, and smoking. The authors reference Beauvais and Segal's 1992 paper "Drug Use Patterns among American Indian and Alaska Native Youth: Special Rural Populations." The Columbia University authors note that "Youths on the reservation were 3.5 times likelier to have tried marijuana, 5.8 times likelier to have tried stimulants and 8.3 times more likely to have tired heroin than were youths in a nationwide sample."

Although no significant differences were seen among adult rural or urban methamphetamine users, rural 10th and 12th graders were substantially more likely to have used methamphetamines within the past year.

Methamphetamine admissions to treatment programs are concentrated in the West and Midwest. Methamphetamine laboratories tend to be located in areas with low population so that the fumes don't attract attention. Initially found more in the West, they are becoming increasingly common in the Midwest, with drug operatives installed among rural laborers. Small labs are also being seen in urban and suburban areas. The manufacturing process is dangerous, with signs of explosions frequently found in the labs. For every pound of the drug produced, five-six pounds of toxic waste are generated.

20. Federal Emergency Management Agency, U.S. Fire Administration. *The Rural Fire Problem in the United States.* Produced under contract EMW-94-C-4443, Aug. 1997, available online at *www.usfa.fema.gov/downloads/pdf/publications/rural.pdf*

This report uses 1993 to 1995 data from the National Fire Incident Reporting System (NFIRS), and 1983 to 1988 mortality data from the National Center for Health Statistics to examine and contrast fires and fire deaths occurring in rural and nonrural areas of the U.S.

The USDA's Rural-Urban Continuum (Beale Codes), was used to define rural for this analysis. Beale Codes 7 ("urban population of 2,500 to 19,999, not adjacent to a metropolitan area," 8 ("completely rural or fewer than 2,500 urban population, adjacent to a metropolitan area") and 9 ("completely rural or fewer than 2,500 urban population, not adjacent to a metropolitan area") were included. In 1993, according to this definition, 19.4 million (7.5 percent of the population) lived in the 45.7 percent of U.S. counties that were considered rural. Unknown information was allocated proportionally throughout the report.

From 1993 to 1995, the incident type of reported fires was similar in rural areas and in the U.S. as a whole. In rural areas, 45 percent of the fires were outside fires, 25 percent involved residential structures, 10 percent involved nonresidential structures, 19 percent were vehicle fires, and 1 percent were other fires. In the U.S. overall, 43 percent of the fires were outside fires, 23 percent involved residential structures, 9 percent involved nonresidential structures, 24 percent were vehicle fires, and 2 percent were other fires.

The profile for civilian fire deaths and injuries also varies little. Residential structure fires caused 69 percent of the rural fire deaths and 60 percent of the rural fire injuries, compared to 72 percent of the U.S. fire deaths and 68 percent of the U.S. fire injuries.

Nonresidential structure fires accounted for 4 percent of the rural fire deaths and 15 percent of the rural fire injuries, compared to 6 percent of total U.S. fire deaths and 13 percent of total U.S. fire injuries. Twelve percent of the North's fires were in or on nonresidential structures, compared to 8 percent in the South.

Vehicle fires accounted for 21 percent of the rural fire deaths and 13 percent of the rural fire injuries, compared to 17 percent of the total U.S. fire deaths and 10 percent of total U.S. fire injuries.

Outside fires accounted for 4 percent of the rural fire deaths and 6 percent of the rural fire injuries, compared to 3 percent of the total U.S. fire deaths and 5 percent of total U.S. fire injuries.

Other fires accounted for 2 percent of the rural fire deaths and 5 percent of the rural fire injuries, compared to 3 percent of the total U.S. fire deaths and 4 percent of total U.S. fire injuries.

Forty-five percent of the rural outside fires were caused by open flame, 16 percent by arson, and nine percent by natural causes. Arson caused 44 percent of the nonrural outside fires. Forty-nine percent of the South's fires occurred outside, compared to 43 percent in the North. When divided by East and West, outside fires accounted for 55 percent of rural fires in the West, but only 36 percent in the East. Although open flame causes more than 40 percent of the rural outside fires in both the East and West, arson caused 29 percent of the Eastern outside rural fires, compared to 12 percent of the Western rural fires.

The 10 percent of rural fires that occurred in or on nonresidential structures accounted for 4 percent of rural civilian fire deaths and 15 percent of rural fire injuries.

Arson and open flame each caused 18 percent of the rural nonresidential structure fires; electrical distribution equipment caused 13 percent. In nonrural areas, arson caused 31 percent of the nonresidential structure fires. Electrical distribution equipment caused 11 percent, and open flame 10 percent. Heating, electrical distribution equipment, open flame and natural causes each caused 17 percent of the rural fatal nonresidential structure fires. In nonrural areas, arson caused 25 percent of the fatal nonresidential structure fires, smoking caused 14 percent, and open flame 13 percent.

The 25 percent of rural fires that occurred in or on residential structures accounted for 69 percent of rural civilian fire deaths and 60 percent of rural fire injuries. Thirty-six percent of these fires were caused by heating, 13 percent by cooking, and 12 percent by electrical distribution equipment. Twenty-six percent of the rural fatal fires were caused by heating, 23 percent by smoking, and 17 percent by electrical distribution equipment. Heating and cooking each caused 23 percent of the rural residential fires with injuries. Children playing, electrical distribution, and smoking each caused 10 percent of these fires with injuries.

One in four nonrural residential structure fires was caused by cooking, 16 percent were caused by heating, and 14 percent were arson. Smoking caused 28 percent of the nonrural, fatal residential fires, 17 percent were arson, and heating caused 12 percent. Cooking caused 30 percent of the nonrural residential fires with injuries; smoking and children playing each caused 12 percent.

Almost three-quarters (73 percent) of rural residential fires occurred in properties without working smoke alarms. In 58 percent, none were present at all. In 15 percent, smoke alarms were present but not operating.

Sixty-five percent of nonrural residential fires occurred in properties without working smoke alarms. In 42 percent, none were present at all, and in 23 percent, smoke alarms were present but not operating.

Twenty-one percent of rural residential structure fires originated in the chimney, 19 percent started in the kitchen, 11 percent started in the lounge, living room, family room, or den, and another 11 percent started in the bedroom.

Flame damage extended to the entire structure in 29 percent of the rural residential structure fires but only 17 percent of nonrural incidents.

Forty-six percent of the rural residential structure fires were extinguished by hose lines preconnected to tanks, 14 percent were self-extinguished, and 12 percent were extinguished by make-shift aids. Thirty percent of the nonrural residential structure fires were extinguished by hose lines pre-connected to tanks, 19 percent were extinguished by makeshift aids, and 17 percent were self-extinguished. Portable extinguishers were used in a larger share of nonrural fires than rural fires.

The time pattern for rural residential fires is consistent with that of the Nation as a whole. The peak time of day for rural fires is between 4 and 8 p.m. However, the peak time for fatal rural residential fires is between midnight and 4 a.m.

Fifteen percent of rural residential fires occur on Saturday and Sunday, with other days of the week having 14 percent each. December (14 percent) is the peak month for rural residential fires, January (12 percent) ranked second, and February (11 percent) ranked third. Winter is the peak time for residential fires nationally.

Fixed stationary heaters, including wood stoves, were involved in 38 percent of the rural residential heating fires. Chimneys (25 percent) ranked second, and fireplaces (11 percent) ranked third. Adhesive, resin, or tar was the type of material first ignited in nearly half of the rural residential heating fires. Sawn wood was first ignited in 19 percent of these fires.

Greater differences were seen in residential structure fire causes between North and South than between East and West. Heating caused 39 percent of the rural residential structure fires in the North but 29 percent in the South. Electrical distribution fires equipment caused 12 percent of these fires in both the North and the South. Cooking caused 11 percent of the rural residential fires in the North and 18 percent in the South. Arson caused 8 percent in the North and 12 percent in the South.

The percentage of rural residential fires originating in the chimney was three times as high in the North as the South. The kitchen was the leading area of origin in the South and the second most frequent area in the North. Bedrooms were the second most frequent area of origin in the South.

Flame damage was more likely to extend to or beyond the structure of origin in rural residential structure fires in the South than in the North. Portable extinguishers were used to put out 11 percent of the rural residential structure fires in the North and 6 percent in the South. Hose lines preconnected to tankers were used in 53 percent of these fires in South and 42 percent in the North.

Smoke alarms were present and operated in 30 percent of the residential structure fires in the North. None were present in 52 percent of these fires. Smoke alarms were present, but did not operate in 18 percent of the incidents.

In the South, smoke alarms were present and operated in only 21 percent of the residential structure fires. None were present in 69 percent of these fires. Smoke alarms were present, but did not operate in 10 percent of the incidents.

The average number of fatalities in rural fatal residential fires was 1.5 for manufactured housing and 1.3 overall. Heating caused 23 percent of the rural manufactured home fires, still the leading cause, but a smaller share of fires than in other rural housing. Electrical distribution equipment caused 19 percent, and cooking 14 percent. Cooking and heating each caused 19 percent of the rural manufactured home fires in the South. In the North, heating caused 26 percent of the rural manufactured home fires, and cooking 11 percent.

Only 25 percent of rural manufactured homes had working smoke alarms. None were present in 63 percent of the fires. Smoke alarms were present but did not operate in 12 percent.

Thirty percent of the manufactured homes in the North had working smoke alarms. None were present in 55 percent of the fires. Smoke alarms were present but did not operate in 15 percent.

In the South, only 19 percent of the manufactured homes had working smoke alarms. No smoke alarms were present in 73 percent of the fires. Smoke alarms were present but did not operate in 8 percent of these incidents.

Flame damage in rural manufactured homes extended to the entire structure 62 percent more often than did flame damage in other rural residential structure fires.

According to mortality data from the National Center for Health Statistics, during 1983 to 1988, an average of 5,764 U.S. fire deaths occurred per year. Only 676 of the victims were rural. White victims accounted for 480 of the rural victims and 3,947 of the victims overall. Little difference is seen in the percentage of fire victims by race or gender in rural versus nonrural areas. The racial picture is different when death rates per million population are considered. During 1993 to 1988, the overall death rate for the U.S as a whole was 23.5 deaths per million. In rural areas, it was 30.9, and nonrural it was 22.8.

The white population had an overall death rate of 19.2 deaths per million. Rural whites had a rate of 24.7 and nonrural whites had a death rate of 18.6.

During this same period, fire killed an average of 167 rural and 1,536 nonrural African-Americans per year. The overall fire death rate for African-Americans was 57.5. In rural areas, the rate was 88.6, the highest of any group studied. The nonrural rate for African-Americans was 55.4.

An average of 28 rural and 38 nonrural Native Americans per year died from fire during this period. The overall death rate for Native Americans was 31.5. The 60.7 rural death rate for this population was more than twice the 23.4 deaths per million population for nonrural Native Americans.

Rural fire victims were slightly more likely to be between 1 and 24 years old, or over 85 than nonrural victims. Rural white victims were more likely to be between 1 and 24 years of age than were nonrural whites. Rural African-American fire victims were somewhat more likely to 75 years of age or older than were their nonrural counterparts. Rural Native American fire victims were more likely than their nonrural counterparts to be under 5 years old.

21. Gomberg, A., and L.P. Clark. *Rural and Non-rural Civilian Residential Fire Fatalities in Twelve States.* NBSIR 82-2519, National Bureau of Standards, Center for Fire Research, Washington: June 1982.

The authors examined residential fire deaths that occurred during calendar year 1978 or July 1978 to June 1979 in 12 States: the high death rate States of Mississippi, Alabama, Oklahoma, Arkansas, Tennessee, and Georgia, and the low death rate States of Connecticut, Utah, Wisconsin, California, Florida, and Delaware. Data from a previous period were used to group the States. (During the year studied, Delaware's death rate was high due to two fires that caused 13 deaths.) A death was considered rural if the fire occurred in an area with less than 2,500 population.

The death rate for rural areas was higher than nonrural areas for both the high and low fire death rates. Fire death rates were markedly lower in the population intervals between 2,500 and 50,000 and those above 500,000. The overall rural fire death rate was 2.5 times the overall nonrural fire death rate.

Rural fires started by heating equipment had a fire death rate of 13.8 deaths per million population, four times the nonrural heating equipment death rate of 3.3. The overall death rate for heating equipment fires was 5.3 deaths per million.

Fires started by smoking materials had an overall fire death rate of 6.1, a rural rate of 7.3, and a nonrural rate of 5.9.

Fires started by cooking had an overall fire death rate of 1.9. The rural rate of 4.2 was twice as high as the overall and three times the nonrural rate of 1.4 deaths per million population.

Electrical distribution equipment fires had a rural death rate of 2.5, more than twice the 1.1 nonrural rate and almost twice the 1.4 overall rate. Open flame caused 1.6 deaths rural deaths per million population, compared to a nonrural rate of 1.0, and an overall death rate of 1.1.

Appliances had a rural death rate of 1.2, twice the nonrural rate of 0.6. The overall rate was 0.7.

The rural rate of 1.1 deaths per million population for fires started by children playing was comparable to the 1.0 rates seen in nonrural areas and overall. Rural areas had a rate of only 0.7 deaths per million population for incendiary and suspicious fires. This is roughly one-third of the nonrural rate of 2.0 and less than half the overall rate of 1.7.

The heating equipment fire death rate was roughly 50 percent higher for rural areas in high death rate States than for rural areas in low death rate States. The latter was still higher than the heating equipment death rate for nonrural areas of high death rate States.

Solid-fueled heating equipment had a rural fire death rate roughly twice that of gas-fueled equipment. The rate for liquid-fueled heating equipment was much lower than either. Fifty-five percent of the deaths resulting from fires started by solid-fueled equipment were in rural areas of high fire death rate States, 21 percent were in rural areas of low fire death rate States, 20 percent were in nonrural areas of high death rate States, and 4 percent occurred in nonrural low death rate States.

The report describes a variety of scenarios. The improper installation of solid-fueled heating equipment (wood stoves, fireplaces, chimneys) caused a rural death rate of 2.43 deaths per million population and a nonrural rate of 0.26. In this scenario, a wood stove may be too close to a wood wall, chimneys or vents may have inadequate clearance from wall coverings or framing, or the floor is not protected from heat or flame.

The rural rate for combustibles (furniture, linens, trash, etc.) placed too close to solid-fueled heating equipment was 0.76/million compared to a nonrural rate of 0.10/million.

Rural States had a death rate of 0.50/million for fires that started when flammable liquids were used to kindle fires in wood stoves and fireplaces. The nonrural rate was 0.02/million. Alcohol was a frequent factor in this scenario.

Ignitions of worn clothing by all types of heating equipment resulted in a rural fire death rate of 2.26/million and a nonrural rate of 0.8/million. Older adults were frequent victims. Victims tended to be sitting close to a wood stove or local heater. In many cases, only the clothing was involved, and the fire department was never notified as victims were transported privately. The rate for these deaths was highest in rural areas in high fire death rate States. Rural areas in low fire death rate States and nonrural areas in high death rate States had similar rates. Gas-fueled equipment was seen as a larger problem in States with high death rates, and solid-fueled equipment was a bigger factor in rural areas.

Portable space heaters that ignited the walls, floors, furniture trash, soft goods, or bedding had a rural fatality rate of 1.66/million compared to 0.48 /million in nonrural areas. These fires resulted when the heater was too close to the items ignited. In many cases, portable heaters were used in place of central heat or fixed local heaters. The death rate from this scenario was highest in rural areas in States with low fire death rates. The ignition of wood paneling by solid and liquid-fueled heaters was a particular problem in rural areas. Fabric ignition by electric heaters was a problem in all areas.

Gas-fueled furnaces and water heaters that ignited wood paneling, walls, ceilings, floors, framing, or insulation in manufactured homes had a rural fatality rate of 0.68 deaths per million residents. None of these deaths were seen in nonrural areas of the States studied. Older manufactured housing, common in rural areas, often lacks fire-rated compartments around heating equipment. Instead, combustible paneling was used.

As noted earlier, rural areas had a cooking fire death rate three times that of nonrural areas. In rural areas, the death rate from unattended cooking (including fires in which the individual was asleep, unconscious, or incapacitated) that ignited cooking materials or walls was 1.18 deaths per million population. In nonrural areas, the rate was 0.28. Alcohol was a factor in about 35 percent the deaths in both rural and nonrural areas. High fire death rate States have high rates of these fatalities in both rural and nonrural areas. The rates are much lower in both rural and nonrural areas of the low death rate States.

Rural areas had a death rate of 0.50/million from fires that occurred when gas or walls were ignited by cooking equipment with installation or maintenance deficiencies (part failure, leak or break, or installation deficiency). None of these fires were seen in the nonrural areas. Leaking liquefied petroleum gas (LPG) was the most common problem.

Rural areas had a death rate of 0.67/million from fires that occurred when cooking equipment ignited misused gases or flammable liquids. (Misuse included unintentional releases, spills, and improper fueling.) The nonrural rate was 0.18/million. In some cases, flammable liquids were used for cleaning or stripping and were ignited by pilot lights or used to kindle wood stoves. The improper fueling or lighting of gas stoves was also a common scenario.

The death rate from fires started by smoking materials was 25 percent higher in rural than nonrural areas. In rural areas, the death rate from residential fires in which mattresses or bedding were ignited by smoking materials (smoking in bed) was 3.20/million compared to 2.18/million in nonrural areas. The rural death rate for upholstered furniture ignited by smoking materials was 3.03/million compared to 2.50/million in nonrural areas. The highest smoking fire death rates were seen in rural areas of low fire death States, followed by nonrural areas in high fire death States and then rural areas of high fire death rate States. The upholstered furniture and smoking fire problem is particularly severe in the rural areas of low fire death States. Intoxication in smoking-related fire deaths was more commonly a factor in high death rate States than the low rate States. Little difference was seen in rural versus nonrural.

Fixed wiring, switches, and receptacles that ignited walls, ceilings, framing, or insulation had a rural fire death rate of 1.42/million compared to 0.32/million for nonrural areas. The problem is more pronounced in rural areas of high fire death rate States. Little difference is seen in rural and nonrural death rates from fires caused by cords, plugs, or bulbs. These are less of a factor in rural areas than the fixed wiring fires.

The death rate from open flame fires is also higher in rural areas, with two scenarios dominating. The death rate from outside open burning (heat of ignition, open fire, match, or lighter) was 0.67/million in rural areas and 0.12/million in nonrural areas. The open burning was generally done for waste disposal of leaves, brush, or trash. Most victims were older men whose clothing ignited while engaged in open burning activities.

The death rate from candle fires started inside was 0.58/million in rural areas and 0.26/million in nonrural areas. The rate is highest in rural areas of high fire death States. Candles were used for light in a notable portion of these incidents. Most victims were juveniles.

Fire death rates for whites and nonwhites were similar in cities with populations of more than 50,000 and in areas in low death rate States with populations between 2,500 and 50,000. The nonwhite fire fatality rate was more than three times as high as that for whites in rural areas of high death rate States, and roughly four times as high in areas with populations of 2,500 to 50,000 in high death rate States. The nonwhite rural fatality rate was roughly 50 percent higher for nonwhites in low fire death rate States.

Sixty-eight percent of the rural fire deaths and 60 percent of the nonrural deaths resulted from fires in one- or two-family dwellings, 25 percent of the rural deaths and 8 percent of the nonrural deaths occurred in manufactured homes, 2 percent of the rural fire deaths and 23 percent of the nonrural deaths occurred in apartments, and 6 percent of the rural and 9 percent of the nonrural fire fatalities occurred in other residential properties.

Smoke alarms were not common during the period of this study. Ninety-four percent of the rural and nonrural deaths occurred in properties without smoke alarms.

Fire departments responded to 94 percent of the residential fire deaths in this study in which response was known. In rural areas of high fire death rate States, nearly 15 percent of the deaths had not been reported to fire departments. Thirty-six percent of the rural fatalities in Arkansas were not reported to the fire service. The largest share of unreported fire deaths were caused by heating equipment fires, often clothing ignitions. Thirty-nine percent of the clothing ignitions in this study were not reported. Some rural areas lack fire department protection.

22. Gunther, Paul. "Rural Fire Deaths: The Role of Climate and Poverty." *Fire Journal* 76, 4 (July 1982): 34-39, 112.

FEMA analyzed death certificate data from the National Center for Health Statistics (NCHS) for fire deaths from 1974 to 1978 in communities with populations under 10,000. (Transportation deaths were not included.) The northern section of the U.S. had moderate to high rural fire death rates. The correlation between rural fire deaths in the North and a) heating degree days and b) freezing days was 0.71 and 0.76, respectively.

Rural fire death rates were low in the central part of the country, and high in the South (North and South Carolina, Georgia, Florida, Tennessee, Alabama, Mississippi, Arkansas, Louisiana, Oklahoma, Texas, New Mexico, Arizona, Kentucky, Virginia, and West Virginia). Alaska and Hawaii were excluded from the analysis. Among the 14 central continental States above 42.5° latitude, rural fire death rates were low in three, intermediate in eight and high in three. Thirteen of the 18 States between 42.5° latitude and the States included in the South had low rural fire death rates, and two were only slightly above the low rate. In the rural South, fire death rates were high in all States except Florida. Rates were very high in Arkansas, Louisiana, Mississippi, Alabama, and South Carolina.

All of the Southern States had high rates of rural poverty. Missouri was the only State outside of the South to have a high rate. Rural poverty rates were either low or intermediate in the remainder. A statistical correlation of 0.79 was found between rural poverty and fire death rates for the national rural population as a whole. The correlation was lower when regions were examined separately, with a correlation of 0.57 in the South and 0.48 in the North. Rural poverty levels vary less in the North. In northern States, the numerical correlation between poverty level and heating degree days was about 0.5.

NFIRS data for rural areas (populations under 10,000) of six Northern States (Maine, New York, Ohio, Oregon, Missouri, and Maryland) in 1977 showed that 38 percent of the fire deaths were caused by heating, compared to 19 percent caused by smoking. Gomberg and Clark's 1982 study of rural fire deaths found that in five northern States (Connecticut, Delaware, Wisconsin, Utah, and California), heating caused 26 percent of the fire deaths and smoking caused 32 percent of the fire deaths in communities with under 10,000 population. Gomberg and Clark also found that heating caused 42 percent of the rural fire deaths and smoking 20 percent in seven Southern States (Alabama, Arkansas, Florida, Georgia, Mississippi, Oklahoma, and Tennessee).

According to data from the 1976 Annual Housing Survey, warm-air furnaces were the primary heat source in 42.5 percent of rural Southern households, 22.3 percent of rural African-American households in the South, and 55.6 percent of households outside the South.

Room heaters without flues were the primary heat source in 15.9 percent of rural Southern households, 34.3 percent of rural African-American households in the South, and 0.6 percent of households outside the South.

Fireplaces, stoves, or portable heaters were the primary heat source in 11.2 percent of rural Southern households, 20.1 percent of rural African-American households in the South, and 4.3 percent of households outside the South.

Room heaters without flues pose a higher risk of asphyxiation. In many cases, installation instructions have not been followed, and many need maintenance. Equipment currently sold is safer.

Bottled gas and wood are used as heating fuels more frequently in the South, with African-American households having the highest usage.

Because of the warmer Southern climate, some may wish to use alternate heat as a means to economize. Smaller homes also may have more combustible materials near the heater, and construction in some cases may be less fire-resistant.

23. Wolf, Alisa. "Fire Safety in the Navajo Nation." *NFPA Journal* 91, 2 (Mar./Apr. 1997): 74-82.

Some homes in the Navajo Nation are more than 2 hours away from the closest fire station and, at the time of the 1990 Census, roughly three-quarters of the Navajo homes did not have telephones. Firefighters must be called individually. No mechanism exists for paging. The first Navajo fire department was funded in 1982.

Some traditional cultural beliefs conflict with firefighting responsibilities. Restrictions and extensive rituals exist related to handling dead bodies. A medicine man's blessing would traditionally be sought before entering a burned building.

According to the 1990 Census, more than half of the Navajo homes used wood for heat. Improperly installed wood stoves caused a substantial share of the winter fires, including a 1996 manufactured home fire that killed a mother and her two children. That the mother, a detention officer, had accompanied the fire chief on a number of fire inspections, that the fire occurred fairly close to a fire station, and that no smoke alarm had been present increased the distress. Officials already had begun working with NFPA's Center for High Risk Outreach, hosting a teacher training workshop for Learn Not to Burn (LNTB)®. The cable channel had shown a video on safe installation of wood stoves 3 days before the fire.

The fire led to the formation of the Navajo Nation Interagency Fire Safety Coalition. The Coalition produced a media campaign and conducted various forms of outreach, including talking to shoppers and meetings at chapter houses. The Navajo Nation Interagency Fire Safety Coalition obtained 1,000 smoke alarms and 300 carbon monoxide alarms. While public educators have been using positive messages for years, this is especially important for Navajo communities. In this tradition, thinking or talking about bad things increases the likelihood they will occur. Success stories were critical.

The risk of fire death among Native Americans (all tribes) throughout the U.S. is two to six times as high as for nonnatives. According to a 1993 report by the Indian Health Services (IHS), fires were the leading cause of unintentional injury death in Native American homes in the U.S. In both Native and nonnative populations, preschool children and older adults faced the highest risk of fire death. According to the 1990 Census, 58 percent of the Native American residents of the Navajo reservation had incomes below the poverty level, and 25 percent were unemployed.

On the reservation, older manufactured homes that predate the 1976 safety standards are common. Often, these homes have only one exit, smaller rooms, and room linings that burn more easily. The Navajo Nation does not have a fire code. A study funded by the Indian Health Service (IHS) found a large share of the smoke alarms on the Devils Lake Sioux Reservation in North Dakota had been disabled because of nuisance alarms.

Wood stoves are not the only cause of heating-related fires on the reservation. Some were started by discarded ashes that had been stored in the home; misuse of propane and kerosene also caused problems. Changes in building materials and practices introduce new risks. Traditional hogans were made of logs and earth and had an opening at the top. An unfortunate side effect of better insulation and tighter construction is an increase in carbon monoxide poisoning.

Because many Navajo children live at boarding schools run by the Bureau of Indian Affairs (BIA) during the week, fire safety training for these children must relate to both life in the dormitories and life at home with their families. Neighbors may live miles away.

The Center for High Risk Outreach also worked with the Ojibwe tribe of Minnesota to adapt the Learn Not to Burn® curriculum to their needs in a program called *Keepers of the Fire.* The Ojibwe worked with NFPA, the Minnesota State Fire Marshal's Office, the Minnesota Safe Kids Coalition, the IHS, and the Red Lake Fire Department on a program that included home inspections. The fire rate fell.

Canada's National Housing Program of the Assembly of First Nations worked with LNTB to develop a curriculum that was more generic than the Ojibwe's that other tribes could adapt.

24. Mobley, Cynthia, Jonathan R. Sugar, Charles Deam, and Lisa Giles. "Prevalence of Risk Factors for Residential Fire and Burn Injuries in an American Indian Community." *Public Health Reports* 109, 5 (Sept./Oct. 1994): 702-705.

In 1992, Mobley, et al., conducted in-person interviews with 68 households in Kitsap County, Washington, that had at least one member of the Suquamish Tribe. The smoke alarm closest to the sleeping area was tested and maximum hot water temperature was measured. The majority of households were on a reservation. Fifteen percent of the households lived in manufactured housing. Seventy-nine percent used electricity as either a primary or secondary heat source. Almost half used wood stoves. Ninety-six percent had at least one smoke alarm, and 95 percent of the installed smoke alarms operated when tested. Fifty-nine percent had experienced an unwanted smoke alarm activation with 84 percent of these caused by cooking.

One or more smokers lived in 59 percent of the households, and one-fourth had someone who smoked in bed. Someone smoked and drank alcohol concurrently in 38 percent of the households.

Nine of the 68 households (13 percent) had experienced a home or yard fire in the previous 3 years. This translates to rate of 6.4 fires per 100 households per year. Nine percent of the households had a member burned or injured by fireworks in the previous 3 years.

25. Kuklinski, Diana M., Lawrence R. Berger, and John R. Weaver. "Smoke Detector Nuisance Alarms: A Field Study in a Native American Community." *NFPA Journal* 90, 5 (Sept./Oct. 1996): 65-72.

In 1995, researchers conducted unannounced visits to 120 homes with at least one Native American in the St. Michaels District of the Devils Lake Sioux Reservation in North Dakota. Smoke alarms were tested and information was collected on alarm type, power source, and power source status. Distances from the devices to possible sources of nuisance alarms were measured. The methodology used to test smoke alarms replicated that used in the Consumer Product Safety Commission's (CPSC's) study.

One third of the households had no smoke alarms at all. Eighty-six percent of the homes owned by Housing and Urban Development had the devices, compared to 46 percent of the privately owned homes. Smoke alarms were found in only 57 percent of the manufactured homes and 69 percent of other single-family dwellings. Sixty-three percent of the homes used natural gas for heat; 15 percent had a wood stove or fireplace. At least one cigarette smoker lived in 73 percent of the homes. Two-thirds had household incomes below the poverty level for a family of four.

A total of 112 smoke alarms were found in 80 homes. Seventy-one percent of the homes with smoke alarms had only one. The vast majority of smoke alarms were ionization. Three were photoelectric and the type was unknown for three. Forty-six percent of the smoke alarms were battery-powered, 44 percent were electrical, and 10 percent were electrical with battery backup.

In 38 percent of the homes with smoke alarms, none were operable. Forty-five percent had at least one smoke alarm that was not working. Overall, 48 percent of the smoke alarms were inoperable. Eighty-six percent had been disabled due to nuisance alarms. Battery-powered smoke alarms were more likely to be disabled due to nuisance alarms than electrical units. The three photoelectric alarms had no history of nuisance activations.

Seventy-nine percent of the households with ionization alarms experienced nuisance activations. Forty-two percent reported more than 25 nuisance activations per unit in the previous year. Cooking was blamed for 77 percent of the nuisance activations; steam from the bathroom was the culprit in 18 percent. Frying was responsible for three-quarters of the cooking nuisance alarms, while baking caused roughly one-third of the cooking activations. The frequency of cooking nuisance alarms decreased as distance from the smoke alarm to the stove increased. Using a stove fan decreased nuisance activations for smoke alarms within 20 feet of the stove, but had no impact when the smoke alarm was further away. No steam activations were reported for smoke alarms more than 10 feet away from the bathroom.

In this study, age of home and smoke alarms, type of heat, insects or debris in smoke alarm, size of home, and power supply were not statistically significant factors in nuisance activations.

26. Struthers, Roxanne and Felicia S. Hodge. "Sacred Tobacco Use in Ojibwe Communities." *Journal of Holistic Nursing* 22, 3 (Sept. 2004): 209-225.

In this study, six Ojibwe spiritual leaders or traditional healers were interviewed about traditional ceremonies of sacred tobacco use and their significance and traditional practices associated with both sacred and commercial tobacco. Three of those interviewed were current smokers, and one had never smoked. They agreed that traditional ceremonial tobacco was made from red willow (kinnikinnick). People can communicate with the Creator by smoking tobacco in a pipe or placing it on the ground near a tree. It is customary to offer tobacco in all ceremonies and many cultural functions. "Deep inhaling was not encouraged because the smoke was not to be enjoyed but was a symbolic gesture meant to cleanse the air, the heart, and the mind." (p. 211) Sacred tobacco is seen as a vital part of the culture.

Commercial cigarette smoking outside of the ritual content was recognized as destructive. However, antismoking or antitobacco messages may be perceived as attacks on the culture. The spiritual leaders and traditional healers can help convey the difference between the two types of uses. Work with this community must include an understanding of, and respect for, the role of sacred tobacco in the community while addressing the very real problem of tobacco abuse and addiction.

27. Vidal-Trecan, G., S. Tcherny-Lessenot, C. Grossin, S. Devaux, M. Pages, J. Laguerre, and D. Wasserman. "Differences between Burns in Rural and in Urban Areas: Implications for Prevention." *Burns* 26 (2000): 351-358.

Nineteen of the 23 French burn units participated in this study of burn victims' characteristics. Senior physicians filled out questionnaires on patients admitted to the burn units from September 1, 1991, through August 31, 1992. Information was captured about demographics, how and where the burn was incurred, the burn itself, and survival. Municipalities with populations under 2,000 were considered rural. Those with populations greater than 2,000 were considered urban. Cases that could not be classified as rural or urban were excluded from this analysis. Thirty-four percent of 1,234 patients resided in rural areas. In rural areas, 28.3 burns were incurred per million population compared to 18.4 in urban areas. More specifically, the burn rate was higher for both rural men and women and for rural children, teens, and young adults than for their urban counterparts. The incidence of occupational burns was twice as high in rural areas.

The rural burn victims tended to be older and less educated. They were more likely to be retired and/or part of a couple than burn victims from urban areas. The only age group with differences in gender patterns based on density was the working adult populations. Seventy-four percent of the rural victims 20 to 64 years of age were male, compared to 66 percent of the urban victims of that age group.

Occupational burns were similar in rural and urban areas. In both settings, the majority of burns occurred indoors, with the most frequent causes being "flames, explosions and electrical installations."

Greater differences were seen in the burns resulting from everyday activities. Rural victims were more likely to be 65 or over, retired, and to have some predisposing factor than were urban victims. Thirty-five percent of everyday rural burns occurred outside compared to 22.5 percent of urban burns. Further breakdown shows that the victims of everyday outdoor burns were more likely to be male, to be either between 20 and 65 or over 65, to have high income, and to be more highly educated than victims of everyday outdoor burns in urban areas.

Although hot liquids were the leading cause of burns in both areas, these burns were less common in rural areas. Larger shares of these burns were seen in urban males and in urban residents at least 65 years of age.

Flames or explosions caused a larger share of burns in rural areas than in urban areas. Rural flame or explosion burn victims were more likely to be male, to be either 20 to 65 or 65 and older, and to have higher income than the urban victims.

Burns caused by open fire and barbecues were more common in rural areas. Rural victims of open fire or barbecue burns were more likely than their urban counterparts to be either 20 to 65 or 65 and over, to have low incomes, and to be less educated.

Rural burns were deeper and more likely to cover more than 10 percent of the body surface area. A larger share of rural burns resulted in death. Rural fatalities were less likely to have had comorbidities.

Only 20 percent of the French population lives in rural areas, but 34 percent of the burn victims were rural. This study dealt only with burns severe enough to be seen in burn units, and excluded minor burns. However, rural residents tend to live a greater distance from these specialty units. The author referenced Thomas et al., who found that some of the difficulties rural residents have in obtaining health care is due to lower socioeconomic and educational levels. The rural population in general tends to be older and have less education. Response and transport times tend to be greater in rural areas.

28. Minnesota Department of Public Safety, State Fire Marshal Division. *Fire in Rural Minnesota—1999,* accessed online at *http://www.dps.State.mn.us/fmarshal/do wnloads/1999RuralFireinMinnesota.pdf* on May 5, 2005.

This analysis examines the fire experience in Minnesota counties with populations under 50,000. Although the USFA used a county population threshold of 20,000 for its

1997 report, *A Profile of the Rural Fire Problem in the United States,* this lower threshold would exclude counties that many Minnesotans consider rural. Statistics are provided for rural and urban counties, as well as the entire State, for 1997, 1998, and 1999.

The breakdown of fire types among outside, residential structure, nonresidential structure is fairly consistent no matter the community size, although nonresidential structure fires were more common in the rural counties. In all 3 years, the percentage of urban fire deaths in residential structure fires was higher for urban counties than rural counties. The percentage of rural fire deaths resulting from vehicle fires was consistently higher than urban fire deaths.

In rural counties, 18 percent of the outside fires were caused by open flames, 11 percent were arson, and 4 percent were started by smoking materials. Arson caused 23 percent of the urban outside fires.

Major differences were seen in causes of rural and urban residential structure fires. In 1999, heating equipment was involved in 22 percent of the rural residential structure fires, but only 9 percent of the urban incidents. Other equipment caused 15 percent of the rural residential fires, but 2 percent of the urban ones. Electrical distribution equipment caused 12 percent of the rural residential fires, and 9 percent of the urban. Cooking caused 11 percent of the rural residential fires, but 24 percent of the urban fires. Open flame caused 10 percent in both areas. Arson caused 8 percent of the rural residential fires, but 15 percent of the urban fires. Seven percent of the rural were residential fires caused by appliances, compared to 10 percent of the urban fires. Smoking caused 5 percent of the rural residential fires and 8 percent of the urban incidents.

Thirty percent of the rural fatal structure fires were caused by furnace malfunctions, and 19 percent were caused by smoking. Forty-three percent of urban fatal residential fires were caused by smoking, and 13 percent were caused by arson.

In 1999, smoke alarms were present and operating in 44 percent of the rural residential structure fires, compared to 47 percent of the urban residential fires. These statistics hide other differences. No smoke alarms were present at all in 33 percent of the rural residential fires and 21 percent of the urban fires. The fire was too small to activate smoke alarms in 3 percent of the rural incidents and 10 percent of the urban ones. Smoke alarms were present and did not operate in 15 percent of the rural residential fires and 20 percent of the urban ones.

Excluding unknowns, flame damage was confined to or within the room of origin in 74 percent of the urban residential structure fires during 1999, but only 58 percent of the rural incidents.

Fireplaces or chimneys were involved in 57 percent of the rural residential heating fires and 49 percent of the urban fires. Adhesive, resin, or tar was the item first ignited in 31 percent of all rural residential heating fires; sawn wood, frequently associated with fireplaces and wood stoves, was first ignited in 23 percent of these incidents.

29. Runyan, Carol W., Shrikant I. Bangdiwala, Mary A. Linzer, Jeffrey J. Sacks and John Butts. "Risk Factors for Residential Fires." *New England Journal of Medicine* 327, 12 (1992): 859-863.

This study compared 151 North Carolina fatal fires from 13 months in 1988 and 1989 with 283 nonfatal, nonchimney home fires with someone home at the time of the fire from the same time period. These fires were said to occur in "predominantly rural areas." Heating equipment was involved in 39 percent of the fatal and 28 percent of the nonfatal fires. Space heaters were involved in 58 percent of the fatal heating equipment fires, and kerosene heaters were involved in 87 percent of the fatal space heater fires. Wood stoves or fireplaces were involved in 45 percent of the nonfatal heating equipment fires; space heaters were involved in 30 percent of these incidents. Kerosene heaters were involved in 55 percent of the nonfatal space heater fires.

Thirty-one percent of the fatal and 6 percent of the nonfatal fires were caused by smoking. Cooking caused 10 percent of the fatal fires and 23 percent of the nonfatal ones. No significant difference was seen between fatal and nonfatal fires in which a) the fire was reported by telephone, b) a 9-1-1 system was or was not present, c) the fire department was or was not comprised entirely of volunteers, and d) the response time was more or less than 5 minutes. Fatality risk was higher for adults 65 or older, people impaired by alcohol or drugs, and for people with physical or mental disabilities.

The risk of fire death was higher in homes without smoke alarms. The absence of these devices had a greater impact when children were present and when no one was disabled or impaired by drugs or alcohol.

Thirty-one percent of the fatal and 21 percent of nonfatal fires occurred in manufactured housing, although manufactured housing accounts for just 11 percent of North Carolina's housing units. The authors noted that "Fires in mobile homes with two or fewer exits were 2.6 times more likely to be associated with fatal than nonfatal fires." (pp. 861-862)

The risk of fire fatality was higher in houses that were at least 20 years old. However, smoke alarms were more likely to be found in homes built after implementation of the State building code in 1976 requiring these devices in new construction. Smoke alarms were found more frequently in manufactured housing than in other housing units and in 54 percent of the owner-occupied units as compared to 18 percent of the rental units.

30. McGwin, Gerald, Victoria Chapman, Matthew Rousculp, John Robison, and Philip Fine. "The Epidemiology of Fire-Related Deaths in Alabama, 1992-1997," *Journal of Burn Care and Rehabilitation* 21, 1, part 1 (2000): 75-83.

In this analysis of Alabama fire deaths, the authors examined a number of factors in the fire deaths investigated by the Alabama State Fire Marshal's Office. A little more than half of the victims over 18 had some alcohol in their blood. They found that fire deaths on the weekend had almost twice the rate of alcohol involvement as deaths during the week. Smoke alarms were present at 42 percent of the urban fire deaths, but only 21 percent of the rural ones. Cigarettes were the leading cause of fires resulting in death.

31. Clark, D.E., C.N. Dainiak, and S. Reeder. "Decreasing Incidence of Burn Injury in a Rural State." *Injury Prevention,* 2000: 6:259-262.

This study examines burn fatality statistics for Maine and the U.S. from 1960 through 1996. Burn discharge data from the Maine Medical Center and other hospitals also are examined. In the sixties and seventies, Maine's death rate from fire or flame averaged about 5.1 per 100,000 population. From 1993 to 1996, it was down to 1.4 per 100,000. The U.S. death rate was 4.2 per 100,000 in 1961 to 1964, and it fell to 1.5 per 100,000 in 1993 to 1996. Maine had a fire or flame death risk compared to the whole U.S. of 1.26 in 1961 to 1964, 1.57 in 1973-1976, and .95 in 1993-1996. Maine burn hospitalizations also declined from 34.8 per 100,000 in the mid-1970s to 10.6 per 100,000 in the mid 1990s.

Single-family wood-frame dwellings are common in Maine. Oil embargoes in 1973 and 1978 resulted in an increase in the number of homes heated by wood. A burn unit was established at Maine Medical Center; burn prevention programs, including smoke alarm programs, were conducted, and schools began paying more attention to burn education in the 1980s. The authors concluded that prevention, particularly the increasing use of smoke alarms and building code improvements, played a larger role in the death rate reduction than did medical care improvements.

32. Wibbenmeyer, Lucy Ann, Margery Josephine Amelon, James Corydon Torner, Gerald Patrick Kealey, Rebecca Marie Loret de Mola, John Lundell, Charles F. Lynch, Thor Aspelund, and Craig Zwerling. "Population-Based Assessment of Burn Injury in Southern Iowa: Identification of Children and Young-Adult At-Risk Groups and Behaviors." *Journal of Burn Care and Rehabilitation* 24, 4 (July/Aug. 2003): 192-201.

Researchers reviewed records of almost 1,400 burn injuries seen in emergency rooms from 1997 to 1999. These emergency rooms were in 10 mostly rural counties in southern Iowa. Burn injuries were more common in the summer. Most injuries were treated at the emergency room and then released. They found that 2.5 percent were admitted, and 3.9 percent were transferred to a burn center. The highest burn rates were seen in children under five and in teens and young adults ages 16-24. Burns occurred at home and at work with close to equal frequency, although this varied by age. Children under five and older adults at least 65 years of age were more likely to have been burned at home. Sixty-one percent of those between 16 and 24 were burned at work.

Burns suffered by children under five and adults over 65 were more likely to have been severe. Severe burns were more likely to have been incurred at home. Two-thirds of the burns were caused by scalds or contact with hot objects; children under five had the highest rate of these burns. Fifteen percent of the burns were flame related. The leading cause of flame-related burns was open fires, with three-fifths resulting from open fires used to burn trash or brush. Accelerants were involved in almost half of the burns caused by open fires, but only 2 percent of burns caused by other types of fires. Trash pick-up is not provided in many parts of the studied region. Open burning is a common means of trash and brush disposal.

"Other burning materials" accounted for 24 percent of the flame-related burns. Work equipment such as torches, welders, or plasma cutters were involved in 44 percent of the burns caused by other burning materials.

Scalds or contact with hot objects caused 90 percent of the burns to children under 5. Scalds and contact burns during food preparation caused 29 percent of the burns incurred by teens and young adults from 16 to 24. Grease was involved in almost half of these food preparation burns. Noncooking work-related injuries, often involving scalds from caustic materials or miscellaneous contact burns, ranked second, welding ranked third, open burning was fourth, and automobile-related burns ranked fifth in this age group.

Open burning was the second leading cause of burns to children between 5 and 15. It is unclear how old children are when they start to perform open burning, but the authors suggest that education programs may be most effective if presented before individuals establish their own habits and practices for open burning.

33. Wibbenmeyer, L.A., M.A. Amelon, R. Loret DeMola, R. Lewis, and G.P. Kealey. "Trash and Brush Burning: An Underappreciated Mechanism of Thermal Injury in a Rural Community." *Journal of Burn Care and Rehabilitation* 24, 2 (Mar./Apr. 2003): 85-89.

From July 1989 to December 2000, 194 (20.1 percent) of people admitted to an Iowa burn unit with flame burns incurred their injuries in the process of burning waste or brush. These burns were eight times as common among males as females. Men between 25 and 44 faced the highest risk of this type of injury, and males between 16 and 24 faced the second highest risk. More than half (53 percent) of these injuries were incurred while burning brush; 29 percent occurred when trash was being burned. Admissions peaked in the summer, followed by the spring and fall. Accelerants were involved in 80 percent of these injuries; gasoline was involved in 90 percent of the accelerant injuries. Ninety percent of the brush-burning burns involved accelerants, compared to 65 percent of the burns from trash burning. Older victims were less likely to use accelerants than younger ones. Six percent of the trash burns resulted from materials in the rubbish that exploded. Six percent of the victims of brush or trash burning died. Elderly victims had the worst outcomes after treatment. Prevention strategies include efforts to reduce the use of accelerants with outdoor burning, and establishment of other means of disposing of brush and trash besides burning.

34. Bailey, Dan W., and Richard E. Montague. "Wildland Fire Management." *NFPA Fire Protection Handbook, 19th Edition,* 2003, Section 7, pp. 95-110.

Between one-fourth and one-third of wildland fires in the 1.5 billion acres of protected U.S. wildlands are incendiary, one-fourth are caused by debris burning, and lightning causes between one-seventh and one-eighth of these fires. The vast majority of these fires are controlled before they reach 100 acres. About 2 percent of wildland fires cause two-thirds of the total burned area. Local fire departments and a number of county, State, and Federal agencies are involved in wildland fire protection. Smokey Bear has been a prevention symbol since 1950. Different techniques and equipment are needed for wildland firefighting than for structural firefighting. Strategies designed for protecting forest resources were not intended for structural protection. As more homes are built in these areas, protection of life and structures requires more attention.

The authors describe three sets of circumstances where structure and wildland fires intersect.

A mixed interface has scattered structures and/or isolated homes in an undeveloped rural area. The risk to individual homes in these areas is high. The occluded interface consists of wildlands, such as a park or conservation land, in an urban area. In a classic interface, a number of homes, such as a subdivision, abut wildlands along a wide front. The classic interface poses a risk of a higher loss of life.

The authors describe the components of wildland fire protection, Firewise communities, fire detection and suppression methods, and issues of topography and fuels.

35. Badger, Stephen G. "Catastrophic Multi-Death Fires of 2003." *NFPA Journal* 92, 5 (Sept./Oct. 2004): 64-73.

An October 2003 California wildland-urban interface fire that spread across 208,000 acres killed 13 civilians and one firefighter. When fatally injured, the victims were attempting to evacuate or to protect property. The cause was not determined.

A separate October 2003 California wildland-urban interface fire that had been intentionally set spread through 91,000 acres and claimed six lives. Again, the victims were either attempting to evacuate or to protect property.

In March of 2003, six people were fatally injured when a legally permitted agricultural fire was set on a sugar cane field. Warnings were given, but the six were either asleep or hiding.

In August of 2003, eight wildland firefighters returning to Oregon from an Idaho fire died after their van collided head-on with a tractor trailer truck and burst into flames. Two died of asphyxiation; traumatic injuries claimed the other six.

36. National Association of State Foresters Core Team. *The Changing Role and Needs of Local, Rural, and Volunteer Fire Departments in Wildland-Urban Interface: Recommended Actions for Implementing the 10-year Comprehensive Strategy—An Assessment and Report to Congress.* Washington, DC, 2003, accessed at *http://www. iafc.org/downloads/Final percent20Rural percent20Fire percent20Report.pdf* on June 1, 2005.

Federal agencies employ fewer than 16,000 full-time and seasonal wildland firefighters. More than 658,000 volunteer firefighters serve in over 24,000 local rural fire departments. These volunteers are often the first to respond to wildland or wildland-urban interface fires.

In 2001, the *Ten-Year Comprehensive Strategy for Reducing Wildland Fire Risks to Communities and the Environment* called for an assessment of the training, equipment, and safety of firefighters who work in the wildland-urban interface. The Core Team involved in this report include representatives from the IAFC, the NVFC, USFA, NFPA, the National Association of State Foresters, the National Association of Counties, the USDA Forest Service, and the Department of the Interior.

The wildland-urban interface requires firefighting equipment, training, and skill in structural and wildland firefighting. Evacuations may be required; communication and interagency coordination are critical. Policies on incorporating local firefighters in multijurisdictional responses inside and outside their immediate areas are required. Plans on responsibility division should be devised before an incident occurs. Radio communication difficulties and incompatibility have been identified as a serious problem. Funding for rural fire departments often is inadequate for their scope of responsibilities. Investments that address these problems can be expected to increase the safety of firefighters and the public, as well as reduce losses and disruption.

37. Nordstrom, David L., Craig Zwerling, Ann M. Stromquist, Leon F. Burmeister, and James A. Merchant. "Epidemiology of Unintentional Adult Injury in a Rural Population." The Journal of Trauma Injury, Infection, and Critical Care 51, 4 (Oct. 2001): 758-766.

More than 1,600 people from a random sample of residents of an all-rural Iowa county were interviewed. The county was considered typical of rural Iowa, with a mix of farms, manufacturing, and services. The sample was stratified by farm, nonfarm, town and nontown, but these categories were self-assigned in the interviews. Forty-two percent of those contacted agreed to a first interview. All individuals over 18 in the household were invited to participate.

Participants were asked about "accidents and injuries" during the previous year. Three criteria were used: 1) any loss of memory or awareness or blacking out; 2) injuries requiring professional attention; or 3) the injury caused restricted activities for 4 hours or more. When more than one injury was recalled, only the most recent was considered. These were subsequently coded using the *International Classification of Disease, 9th Revision, Clinical Modification.*

Participants were also asked about demographic data, and behavioral and social factors. Positive answers to two of four questions from a diagnostic tool for alcohol abuse (CAGE) were considered indicative of alcohol abuse. A positive response to a question about five or more drinks in the past 30 days was considered indicative of binge drinking. Depressive symptoms were measured by an abbreviated version of the Center for Epidemiologic Studies Depression (CES-D) scale. Participants tallied positive answers to

10 questions from the National Comorbidity Study to assess antisocial personality traits. Four or more positive answers suggested antisocial personality.

Twenty-three percent of the people reported an injury in the past year. Roughly one-third (35.4 percent) were sprains or strains, and one-fifth (18.9 percent) were wounds or lacerations. Burns were the diagnosis in 3.4 percent of the injuries. Health care was received for four-fifths of the injuries. One-quarter (25.5 percent) of the injuries were caused by overexertion and 21.6 percent were caused by falls. Fires or burns caused 2.1 percent of the injuries. The percent reporting injuries decreased with age.

Women who scored high on the depression scale were more likely than other women to have suffered an injury in the past year. Men who were dependent on or abused alcohol were more likely to have had an injury than other men. The authors noted that in their study, 48.5 percent of the injuries occurred while the individual was working, compared to 26.3 percent found nationally by the U.S. National Health Interview Survey ((NHIS).

38. Baker, Susan P., R. A. Whitfield, and Brian O'Neill. "County Mapping of Injury Mortality." *Journal of Trauma* 28, 6 (1988): 741-745.

The authors mapped injury death rates by county for all unintentional injury deaths, firearm homicides, suicides, house fires, drownings, and drowning deaths of children under 5 years old. The International Classification of Disease E-Codes were used with the National Center for Health Statistics death data for 1979 to 1981. The authors noted that death rates can vary widely within States. A State may have a large city with a high death rate but the rate may be low in its suburban and rural counties. High death rates in less populated areas may be obscured if the rate is lower in denser areas.

They found that the West and the South had the highest death rate from unintentional injures, with the highest rates in rural counties. The rate is much lower in the Northeast.

Nevada, New Mexico, and Wyoming have high suicide rates in roughly half of their counties. These rates are also elevated in parts of Florida and in many Appalachian Mountain Counties.

Firearms caused two-thirds of all homicides. Firearm homicide rates are high in low income areas. This is particularly true for the South and large cities. Ten sparsely populated, low-income New Mexico counties also had high firearm homicide rates.

The South, with the exception of Florida, had very high rates of deaths from house fires, with especially severe problems in Mississippi River Valley and Atlantic coastal low-income counties. House fire death rates were generally low in the Western part of the country. The authors note that noncentral heat is more common in low income areas of the South, more older wooden buildings are found in the South than in most parts of the country, and distances are greater for rural firefighters.

Drowning was the one factor that correlated with high income. Child drowning rates were high in areas where family swimming pools are common.

39. Bull, C. Neil, and Shari DeCroix Bane. "Program Development and Innovation." *Journal of Applied Gerontology* 20, 2 (June 2001): 184-194.

The term "rural" describes a variety of settings, including "frontier" counties with populations less than six per square mile and rural counties that abut urban ones. Differences also exist between farm and nonfarm rural populations. Some of the issues involved in developing and providing mental health services in a rural area are addressed.

Four core issues are discussed: geographic isolation, economic deprivation, human service infrastructure, and economies of scale. The distance that must be traveled is a part of the isolation. Public transportation is close to nonexistent. Older adults may feel that giving up driving is not an option due to a lack of alternatives. Terrain and weather can also make driving difficult. Costs for long-distance calls, fuel, and travel time add up quickly.

Economic deprivation is exacerbated by the tendency for many rural areas to rely heavily on one industry, activity, or service for local livelihoods. Economic shifts or plant closures can be devastating to residents' income and local tax revenues. Some Federal and State programs mistakenly assume that services can be provided at less cost in rural areas, and do not fund adequately. Because rural incomes are lower and fewer foundations are rural, there are fewer charitable resources available for programs or for the matching funds necessary to qualify for some grants.

The human service infrastructure has experienced consolidations and closings. There is a shortage of technical equipment and skilled personnel. Rural youth often move away and more women are working, reducing the volunteer pool that might partially alleviate the lack of paid workers.

Economies of scale are difficult because the numbers of people and suppliers are simply not there. In some cases, there is a sole supplier. Competitive bidding may not be possible.

Rural housing is said to be less well maintained, rural residents less educated, and nonfarm rural residents tend to be in poorer health.

The independence associated with rural life, particularly among the elders, often results in a resistance to using or accepting services or assistance. Seven points are made about transferring urban programs to rural areas:

1. Expectations may need to be scaled, back, particularly if success is defined as number of people served.

2. It is often necessary to scale services to offer only the highest priority (as defined by the community), rather than offering the full range.

3. Program duplication should be avoided and offerings coordinated so that each agency offers programs it can do best.

4. Rules and regulations should be handled with some flexibility as bookkeepers and accountants tend to be in short supply. Budget waivers should be sought

when expenses will be higher than expected for items like long-distance calls and mileage.

5. Do not expect economies of scale or more than one provider bidding.

6. Create partnerships or reciprocal agreements so that the jurisdictional or administrative boundaries do not interfere with services.

7. Plan for challenges in recruiting and keeping qualified personnel. Hire people with multiple competencies rather narrow specialists.

Many services are delivered without benefit of formal office space. These services may be delivered from stores, churches, restaurants, or vehicles. Gatekeepers, including mail carriers, beauticians, and neighbors, can be used for referrals. The cooperative extension network is recommended as a vehicle for educational programs. Aging, nutrition, and hospital programs can support and publicize a new endeavor. Programs that operate in isolation are less likely to be successful.

40. DeCroix Bane, Shari, and C. Neil Bull. "Innovative Rural Mental Health Service Delivery for Rural Elders." *Journal of Applied Gerontology* 20, 2 (June 2001): 230-240.

Distance, isolation, and resource shortages interfere with both problem identification and obtaining help with mental health issues. Ideally, an informal support system of family, neighbors, and friends helps rural elders through predictable life crises. Most services wait to be contacted and traditional outreach tends to find those who are functioning fairly well. Several components common to successful direct service and educational programs are discussed.

In many programs, one individual's enthusiasm, work, and commitment were critical in organizing and persuading others to establish a program. A credible "natural leader" from the community knows how to present the concept in a way that would be acceptable locally and could motivate other groups to participate.

Services had to feel comfortable to clients. Impressions of comfort can be based on the program's appearance, location, time, and expense. Some programs provided services through nontraditional but trusted partners such as grain dealers, banks, and utility providers. Flexibility to choose to use portions of the services **when** they choose was important. Rural elder women tended to be cautious about accepting formal services and were hesitant to accept when they felt that they could not reciprocate.

Two direct service models were described: gatekeepers and peer counseling. Gatekeepers routinely have contact with people who themselves would not seek services. These gatekeepers make referrals to appropriate agencies. Elder counseling uses trained elder volunteers from the same community as part of the mental health team. Elders were more open to peers than they were to professionals whom elders feared might threaten independence. Peers often have a more realistic understanding of the client's situation.

Three educational program models that trained nonprofessionals to recognize mental health problems and make appropriate referrals were also discussed. The nonprofessionals received cross-training about a number of organizations. The organizations had to gain a better understanding of the mental health needs of rural older adults and to collaborate in service provision.

A Missouri project trained trainers, using the well-established networks of area agencies on aging and university extensions. At eight sites in the State, local committees comprised of representatives of these two organizations, and community leaders, including clergy, community mental health providers, elder volunteers, or senior center directors, coordinated recruiting of people who wanted training in mental health and aging. The model a) improved the providers' ability to recognize mental health needs, b) identified rural older adults who needed service, c) provided information on referral programs, d) assisted providers in notifying possible clients about their services, and e) helped providers inform agencies about possible clients.

An Arizona program presented workshops for rural health (not mental health) staff, volunteers, family and community members on identifying signs of mental illness in older adults, how to respond more effectively to people with these problems, and appropriate referrals. Culturally specific curricula were developed for rural Anglos, Latinos, and Native Americans. The local supports helped continue training after the program ended.

A Pennsylvania program used a cross-system training model and focus groups to create a curriculum. Volunteers, gatekeepers, and nonprofessional caregivers were trained to give educational presentations in their communities. Recipients of training then conducted more training. This approach brought staff of the mental health and aging systems together at meetings.

All three programs used committees and focus groups with community leaders in formal and informal roles. The committees helped develop the curricula specific to their locations and played key roles in marketing the program and recruiting people to participate.

The importance of rural-specific material was stressed. Crisis intervention materials that advise calling 9-1-1 are not appropriate in areas that lack that service. The sheriff may be the emergency contact.

41. USDA Rural Development Housing Programs." accessed online at *http://www. rurdev.usda.gov/rhs/* on May 5, 2005.

The Home Repair Loan and Grant Program (Section 504) offers loans of up to $20,000, payable over 20 years, at 1 percent interest, and for people who are at least 62 and unable to repay a loan, grants of up to $7,500 for repairs or to remove health or safety hazards. Loans may also be used for improvements or modernization. Loans and grants may be combined in certain circumstances. Funds can be used for roof repairs, to install central heat or running water, or to put in a wheelchair ramp. Properties that receive these funds are not required to meet the requirements of Rural Health Services codes and, as long as major safety and health hazards are removed, it is not essential to remove all hazards. Water and water systems must conform to local health requirements and usually will be expected to meet the requirements of the Rural Health Services.

Through its community facilities program, the Rural Emergency Responders Initiative offers financial assistance for equipment, vehicles, and/or buildings for fire, police, heath care, and other activities. Priority for grants is given to low-income communities and communities with populations under 5,000. Health care, public safety, and community or public service projects also receive priority. Loan programs are also available for rural areas and small towns with populations up to 20,000.

42. Fire Service Training Services, Alaska State Fire Marshal's Office. "Project Code Red" accessed online at *http://www.dps.state.ak.us/fire/asp/pcr.asp* on June 1, 2005.

Senator Stevens and the Alaska State Fire Marshal's Office led the development of the Micro-Rural Department to address the fire problem and lack of firefighting equipment in the small rural areas of Alaska. Using new and existing technologies and State-certified training to cope with the extreme winter temperatures, lack of hydrants, and, in many cases, roads, the nonprofits Alaska Rural Partners (ARP) and Alaska Village Initiatives, Inc. (AVI) developed the Micro Rural Fire Department, or Project Code Red, in 2001. Additional support came from the U.S. Bureau of Land Management, U.S. Department of the Interior, Bureau of Indian Affairs, the Alaska Federation of Natives, Alaska Village Electrical Coop, Alaska Rural Development Council, ANCSA Regional and Village Corporations, and 86 Alaskan Villages.

This project provides 600 gallons of environmentally safe foam for firefighting. The unit can be recharged in less than 5 minutes for under $50. The equipment allows firefighters to fight a fire from 75 feet. The program equips five firefighters with fully supplied trailers that can be transported by all-terrain vehicles, snow machines, pick-up trucks or by hand, even on boardwalks and trails. The equipment has been tested by Underwriters Laboratories down to temperatures of -40 °F (-40 °C). The trailer is shipped in a heated and insulated container that doubles as a firehouse. The total cost is about 70 percent less than the cost of a new fire engine.

State-certified firefighter training, based on an adapted version of NFPA Firefighter I, is also included. The training is intended for fire departments without protective gear, that have only a limited water supply, and may have only portable extinguishers and pumps.

43. Boddie, Stephanie C. "Fruitful Partnerships in a Rural African-American Community: Important Lessons for Faith-Based Initiatives." *Journal of Applied Behavioral Science* 38, 3 (Sept. 2002): 317-333.

This article begins with a review of the role of faith-based communities in providing social services and the characteristics of programs supported by these congregations, particularly those supported by rural African-American congregations. Historically, many congregations have provided some type of social services, including visits to the sick and other health programs, international relief, educational or cultural activities, environmental issues, and food distribution. Chaves and Higgins (1992) found that African-American congregations were more likely than white congregations to be involved in civil rights and providing basic needs to the immediate community. Rural churches are described as similar to an extended family. Earlier studies generally found that, due to a lack of resources, rural African-American churches were less likely to provide services for older adults, to collaborate with other agencies, to provide day care, substance abuse programs, or assistance with basic needs such as food, clothing, or housing. However, these rural churches are important forces in leadership development and community organizing. They also maintain strong interactions with the private sector. In both rural and urban African-American communities, the churches are among the chief sources of influence and support. More recent studies have found that rural African-American congregations are involved in a wide range of social services and educational programs.

The author conducted an indepth study of the faith communities and their services in Boley, Oklahoma. Boley is a historically African-American town settled in response to racial discrimination and oppression. Boley remains largely African-American. Four hundred people live in the town proper, and over 900 live in the surrounding area. The average church in the community has 22 active members, with roughly one-third of the members 65 or older, and 39 percent 25 to 64. Three out of four congregants have annual household incomes of under $25,000; one-fourth of the annual incomes are between $25,000 and $50,000 per year. The churches are described as the "most prominent and well-maintained structures in town." (p. 325) The churches often provide members, leadership, and resources to social programs independent of government policies. These churches sponsor a wide variety of programs such as literacy, drug and alcohol prevention, and youth outreach, and support many other programs such as the library, Scouting, and youth recreation. Eight churches that cannot provide formal services do provide volunteers or financial support for other programs within the community. Fourteen of Boley's congregations are members of a ministerial alliance, and two of the three remaining assist when called upon. The churches interact with business to provide services. The alliance "has been most effective in maintaining and institutionalizing the programs initiated by the various participating churches, including a community choir, a senior center, funeral ushers, a crime watch group, financial assistance, historic preservation, hosting of holiday celebrations and civic group meetings, sponsorship of the Sunday School Institute, vocational training, a volunteer fire department and a literacy program that culminates in a GED." (p. 328) By working with other congregations, even the smallest become important parts of the social service resources. The 15 bivocational ministers donated time and brought skills and networks from their other careers to social service provision.

44. Fries, Elizabeth A., Jennifer S. Ripley, Melissa I. Figueiredo, and Beti Thompson. "Can Community Organization Strategies be Used to Implement Smoking and Dietary Changes in a Rural Manufacturing Work Site?" *The Journal of Rural Health* 15, 4 (Fall 1999): 413-420.

The health of rural Americans is not as good as that of other Americans. The death rate from preventable cancers is unusually high in the rural South. The rural South has larger percentages of poor, illiterate, elderly, and minority residents. Fewer prevention activities are conducted in rural areas. The authors describe a pilot project conducted at a multishift manufacturing plant in Mecklenburg County, Virginia. Census data identified the county as 80 percent rural, and the population was described as "underserved in terms of medical care, patient education, and cancer education." (p. 414) County population was estimated at 45 percent African-American, illiteracy was believed to exceed 25 percent, and income was below average for Virginia and the country. Manufacturing, industry, and tobacco farming play major roles in the local economy. Four manufacturers were considered for the program. The site that was chosen was operating around the clock and had a reputation as a well-run company with a stable work force. Eighty-five percent of the employees were involved in manufacturing, and 15 percent were managerial or clerical. Most employees at this site were African-American and had a high school education. Management interest was a factor in site selection.

The project began with the recruitment of a Health Advisory Board (HAB). The company's personnel director assisted with this task. Supervisors gave all employees an invitation to an introductory meeting. A dozen volunteered at the outset to be on the HAB. The HAB was assigned to "identify and develop activities and health promotion ideas to address cancer prevention and changing diet and smoking among their co-workers." (p. 415). Project staff attended the monthly meetings at the site.

With support from the project, the HAB conducted one activity per month over 9 months, usually around shift changes. By the second month, they also had posted relevant posters in eating areas and areas of egress. Short surveys were completed before and after the program regarding demographics, smoking behavior, diet, workplace norms for smoking, and, for smokers, readiness to quit smoking. Completion rate was high.

Some information (generally printed) was given to all employees. Thirty-five percent attended a dietary fat screening. Before the program began, 40 percent of the survey respondents were smokers, compared to 35 percent at the end. Only four percent attended a nutrition seminar that was offered twice. A "stop smoking" contest attracted 13, or 18 percent, of the known workplace smokers. All 13 were able to quit for 24 hours; five were still not smoking after 2 months. In addition, smokers at the plant tend to be somewhat readier to quit smoking at the end of the program than they were at the outset. Daily consumption of fruit and vegetables increased, and employees became more self-confident in reducing fat consumption.

The total program, including refreshments for monthly HAB meetings, pamphlets and posters, smoking cessation supplies, training on and administration of fat analysis tests, a nutrition speaker's fee, and $100 per day for a health educator for 11 days, came to about $1,925. This does not include donations of pamphlets, posters, and prizes for the "stop smoking" contest or the time spent by employees on this project.

The authors conclude that community organizing strategies at the workplace may be appropriate to reach minority rural residents. "A low-intensity community organizing approach with minimal intervention resources can reach employees in such work sites and produce small behavioral and attitudinal changes." They recommend future randomized trials.

45. Dorresteyn-Stevens, Christine. "The Rural Hospital as a Provider of Health Promotion Programs." *Journal of Rural Health* 9, 1 (Winter 1993): 63-67.

A 1984 American Hospital Association national survey of hospitals found that 85 percent provided health promotion programming. Different studies gave different impressions on the prevalence of this programming in smaller hospitals.

The 128 North Carolina community hospitals, including 45 with fewer than 100 beds, were surveyed about their health promotion programs. Almost all of the 45 small hospitals served rural populations. The study used Pack et al.'s definition of health promotion as "any attempt to influence individuals to adopt and maintain healthy lifestyles, frequently requiring the relinquishment of unhealthy behaviors." (p. 64).

Sixty-three percent of all hospitals and 64 percent of the small hospitals responded. Ninety-three percent of the rural hospitals, and 96 percent overall, offered at least one health promotion program. Health promotion programs were equally likely at hospitals regardless of hospital size.

At least half of the responding rural hospitals offered first aid and cardiopulmonary resuscitation, AIDS education, nutrition prenatal education, and breast self-examination. Other common programs included smoking cessation, weight control, and stress management. Hospital employees were the most common target audience, with non-patients and members of the community ranking second. Some programs were provided for inpatients and outpatients. The target audience varied by program.

Forty-eight percent of the rural hospitals charged program participants a small fee. Thirty percent of the rural hospitals did not charge, but instead counted the costs as part of the general budget. Fourteen percent had the cost included in the daily room charge. Eighteen percent of the hospitals passed the entire cost on to participants. Combinations of financing methods were used by some hospitals. None used grants to fund these programs.

The "inservice" department was the coordinating department for health promotion programming in one-third of the small hospitals. Nursing departments and "more than one department" were each cited in one-quarter of the hospitals. Inservice departments are usually run by nurses. Nurses were the primary coordinators of this programming at 85 percent of the small hospitals.

Health promotion programming was sometimes considered part of public relations. Programs targeting hospital employees may be developed for certification, to reduce absenteeism, or to increase productivity. They may also be tested on employees before being shared with the community. Most program sessions were conducted by regular staff. Having programs generally coordinated through nursing or inservice departments increases the likelihood that these programs will be for hospital staff and may make the development of programs for the community less likely.

Hospitals could serve as links between health care, businesses, and community agencies. These institutions can be key players in establishing a group to coordinate health promotion activities in the area. The inclusion of community residents can increase participation and better identify the priorities of the particular community. Coordination is essential in rural areas for optimal use of scarce resources, to prevent program duplication and provide varied programs to address different topics and needs. Information should be communicated in many different ways, including posters at stores, fliers, and media. Programs should be held at sites that area familiar and accessible, including stores, schools, and places of worship.

Appendix B. *Rural Highlights of U.S. Fire Service Needs Assessment and Volunteer Firefighter Onduty Fatalities*

Rural Fire Service Needs Assessment

The material that follows presents results based on data from U.S. local fire departments participating in a needs assessment survey. The full national results, covering all community sizes, were published as DHS/FEMA/USFA report FA-240 in December 2002. NFPA published State reports, with selected results for each State, in 2004. Results of a second needs assessment survey were published in October 2006 as DHS report FA-303.

In this report on rural highlights, we use the U.S. Census Bureau definition of a rural community as a community with less than 2,500 population. Some rural programs include communities as large as 9,999 population. The original needs assessment report contains results for 2,500 to 4,999 population and from 5,000 to 9,999 population. See the survey methodology section at the end for a more detailed discussion of the statistical methodology used.

The questionnaire principally involved multiple approaches to answering the question "what does the fire department need?". Most of the questions were intended to determine what fire departments **have**, in a form that could be compared to existing standards or formulas that set out what fire departments **should** have. Some of the questions asked what fire departments have with respect to certain cutting-edge technologies for which no standards yet exist, and no determinations of need have yet been proposed.

The questionnaire also sought to define the emergency-response tasks that fire departments considered to be within their scope. For such tasks the survey asked how far departments would have to go to obtain the resources necessary to address those tasks or an illustrative incident of that type. Clearly, if departments believe the resources they would need are only available from sources separated from them by great distance—and the associated likelihood of significant delay in attaining those resources, there may be a need for planning, training, or arrangements for equipment that can be more quickly accessed and deployed, to assure timely and effective response.

Glossary

Here are standard definitions for some of the specialized terms used in this report:

Advanced Life Support. Functional provision of advanced airway management, including intubation, advanced cardiac monitoring, manual defibrillation, establishment and maintenance of intravenous access, and drug therapy. [from NFPA 1710, *Standard for the Organization and Deployment of Fire Suppression Operations,*

Emergency Medical Operations, and Special Operations to the Public by Career Fire Departments, 2001 edition.]

Basic Life Support. Functional provision of patient assessment, including basic airway management; oxygen therapy; stabilization of spinal, musculoskeletal, soft tissue, and shock injuries; stabilization of bleeding; and stabilization and intervention for sudden illness, poisoning and heat/cold injuries, childbirth, CPR, and automatic external defibrillator (AED) capability. [from NFPA 1710, *Standard for the Organization and Deployment of Fire Suppression Operations, Emergency Medical Operations, and Special Operations to the Public by Career Fire Departments*, 2001 edition.]

Emergency Medical Care. The provision of treatment to patients, including first aid, cardiopulmonary resuscitation (CPR), basic life support (EMT level), advanced life support (Paramedic level), and other medical procedures that occur prior to arrival at a hospital or other health care facility. [from NFPA 1581, *Standard on Fire Department Infection Control Program*, 2000 edition] In this report, reference is made to "EMS" or "emergency medical service," which is the service of providing emergency medical care.

First Responder (EMS). Functional provision of initial assessment (i.e., airway, breathing, and circulatory systems) and basic first-aid intervention, including CPR and automatic external defibrillator (AED) capability. [from NFPA 1710, *Standard for the Organization and Deployment of Fire Suppression Operations, Emergency Medical Operations, and Special Operations to the Public by Career Fire Departments*, 2001 edition.]

Hazardous Material. A substance that presents an unusual danger to persons due to properties of toxicity, chemical reactivity, or decomposition, corrosivity, explosion or detonation, etiological hazards, or similar properties. [from NFPA 1500, *Standard on Fire Department Occupational Safety and Health Program*, 1997 edition.]

Structural Firefighting. The activities of rescue, fire suppression, and property conservation in buildings, enclosed structures, aircraft interiors, vehicles, vessels, aircraft, or like properties that are involved in a fire or emergency situation. [from NFPA 1500, *Standard on Fire Department Occupational Safety and Health Program*, 1997 edition.]

Technical Rescue. The application of special knowledge, skills, and equipment to safely resolve unique and/or complex rescue situations. [from NFPA 1670, *Standard on Operations and Training for Technical Search and Rescue Incidents*, 1999 edition.]

Wildland/Urban Interface. The line, area, or zone where structures and other human development meet or intermingle with undeveloped wildland or vegetative fuels. [from NFPA 295, *Standard for Wildfire Control*, 1998 edition.]

The U.S. Fire Service

Career and volunteer fire departments

Most U.S. fire departments are volunteer fire departments, but most of the U.S. is protected by career firefighters. Three of every four U.S. fire departments are all-volunteer fire departments, but only one of every four U.S. residents are protected by such a department. Only one in 17 fire departments is all-career, but two of every five U.S. residents is protected by such a department. Fire departments split roughly 9-to-1 between the all- or mostly volunteer departments versus the all- or mostly career departments, but the population protected splits roughly 2-to-3 the other way.

Volunteers are concentrated in rural communities, while career firefighters are found disproportionately in large communities. Rural communities, defined by the U.S. Bureau of Census as a community with less than 2,500 population, are 99.5 percent protected by all- or mostly volunteer departments and account for just over half of all the all- or mostly volunteer departments in the US.

In rural communities, there are a total of 13,440 fire departments, of which 43 (0.3 percent) are all-career, 32 (0.2 percent) are mostly career, 454 (3.4 percent) are mostly-volunteer, and 12,911 (96.1 percent) are all-volunteer.

Community size is related to the U.S. fire service not only in terms of the relative emphasis on career versus volunteer firefighters but also in terms of the challenges faced by local departments. However, it is possible to exaggerate those differences. Even a rural community can have a large factory complex, a large stadium, or even a highrise building, with all the technical complexities and potential for high concentration of people or valued property that such a property entails. It is likely that every fire department will need to have some familiarity with every type of fire and every type of emergency, if not as part of protecting their own community, then at least in their role as a source of mutual aid or a component of regional or even national response to a major incident.

In any community, fire burns the same way in open or in enclosed spaces. Fire harms people and property in the same ways. And the resources and best practices required to address the fire problem—or any other major emergency—safely tend to be the same everywhere. What may differ is the defined scope of responsibility of the local fire department and the quality and quantity of resources available to the department to perform those responsibilities.

Budgets and revenue sources

The first questions of the Needs Assessment Survey focused on big-picture budget and revenue issues. Nearly twenty-one percent of departments (20.7) in rural communities, (compared to 35.2 percent of all departments) said they had a plan for apparatus replacement on a regular schedule. This is the kind of long-range, capital-budget type of plan that might be more likely in a community with established, institutionalized sources of revenue for the fire department, as one would expect to see with a career fire department.

In rural communities, 12.0 percent of departments (compared to 21.7 percent of all departments) say their normal budget covers the costs of apparatus replacement. The rest must seek funds in a more ad hoc fashion, such as seeking a special appropriation for such a purchase.

Most revenues for all- or mostly volunteer departments in rural communities are covered by taxes, either a special fire district tax or some other tax (62.6 percent of departments). Other governmental payments include reimbursements on a per-call basis (used by 1.9 percent of departments), other local government payments (4.8 percent), and State government payments (6.6 percent). Fundraising was used by 19.1 percent of departments and other sources by 5.0 percent of departments.

Smaller communities, with less certain sources of revenue, are more likely to obtain their apparatus either used or converted from a nonfire-department design and use. Vehicles that were purchased or, less often, donated used accounted for an average of 42 percent of apparatus for departments protecting communities with less than 2,500 population (34.6 percent purchased used, 7.0 percent donated used). Converted vehicles accounted for an average of 15.8 percent of apparatus for departments protecting communities with less than 2,500 population. Averages for other arrangements were 39.9 percent for purchased new, 0.7 percent for donated new, and 1.6 percent for other.

Because converted vehicles were not originally designed for fire department use, it can be especially challenging to ensure that they are safe and effective, but it is essential that any vehicle, converted or not, be evaluated for its compliance with applicable standards, in order to avoid undue hazard or risk to the firefighters who operate it. A starting point for such an evaluation can be NFPA 1912, *Standard for Fire Apparatus Refurbishing.*

Personnel and Their Capabilities

Number of firefighters

Table A indicates the number of career, volunteer, and total firefighters, by the size of the community their fire department protects. These numbers will be used repeatedly throughout the report to convert survey responses phrased in terms of the fraction of a department's firefighters having a characteristic into estimates of the number of firefighters having that characteristic.

Table A. Number of Career, Volunteer, and Total Firefighters by Size of Community

Population Protected	Career Firefighters	Volunteer Firefighters	Total Firefighters
1,000,000 or more	32,700	150	32,850
500,000 to 999,999	28,400	4,900	33,300
250,000 to 499,999	26,600	4,250	30,850
100,000 to 249,999	39,750	8,550	48,300
50,000 to 99,999	37,750	11,000	48,750
25,000 to 49,999	40,000	29,300	69,300
10,000 to 24,999	38,850	86,050	124,900
5,000 to 9,999	12,200	112,300	124,500
2,500 to 4,999	5,050	157,600	162,650
Under 2,500	4,800	408,750	413,550
Total	**266,100**	**822,850**	**1,088,950**

The above projections are based on 8,012 departments reporting on Questions 7 and 8. Numbers are estimated to the nearest 50 and may not add to totals due to rounding.

Q. 1: Population (number of permanent residents) your department has primary responsibility to protect (excluding mutual-aid areas)
Q. 7: Total number of full-time (career) uniformed firefighters   Q. 8: Total number of active part-time (call or volunteer) firefighters

Adequacy of number of firefighters responding

NFPA 1720, *Standard for the Organization and Deployment of Fire Suppression Operations, Emergency Medical Operations, and Special Operations to the Public by Volunteer Fire Departments,* calls for a minimum of four firefighters on site before an interior attack on a structure fire is begun. There are difficulties in applying these standards to the statistics collected. Responding career firefighters from mostly volunteer departments are not shown, the statistics obtained are average numbers responding rather than minimum numbers responding, and the threshold number of four is combined with averages from three to four in the questionnaire. Nevertheless, some limited observations are possible.

Departments that deliver an average of one or two volunteers to a midday house fire almost certainly fall below the minimum of 4 firefighters in most responses for departments protecting communities with less than 2,500 population, because those departments average only one or two career firefighters on duty for the department. Departments that deliver an average of one or two volunteers (and an unknown number of career firefighters) to a midday house fire constituted 3.0 percent of departments protecting communities with less than 2,500 population.

Departments protecting communities of less than 2,500 population that deliver an average of three or four volunteers may fall below the minimum number of four firefighters in some responses. Departments that deliver an average of four or fewer volunteers to a midday house fire constituted 21 percent of departments protecting communities with less than 2,500 population.

Most of the all-volunteer or mostly volunteer fire departments averaging fewer than four firefighters responding to a midday house fire, and therefore often failing to achieve the minimum standard response to initiate an interior attack, are departments protecting communities with less than 2,500 population. Because roughly one-fourth of the U.S. resident population live in communities of this size, this suggests roughly 5 percent of the U.S. population is protected by fire departments that average fewer than four firefighters responding to a mid-day house fire and so may often fail to achieve the minimum standard response to initiate an interior attack. (The 5 percent is calculated as one-fourth of 21 percent.)

If this is translated into firefighters, then 21 percent of volunteer firefighters serving communities with less than 2,500 population translates into at least 86,000 volunteer firefighters serving in fire departments where the achievement of a standard minimum response to a midday house fire is problematic.

Extent of training and certification, by type of duty

Structural firefighting

Only 1.5 percent of fire departments protecting rural communities say structural firefighting is **not** within the scope of the fire department.

The survey asked how many of the personnel responsible for structural firefighting have received formal training. Answers were solicited in the form of: All, Most, Some, and None. For analysis purposes, "Most" was estimated as two-thirds and "Some" was estimated as one-third. Based on these assumptions, 151,000 firefighters serving rural communities (37 percent of total firefighters serving these communities) are estimated to need formal training because they work in departments with responsibility for structural firefighting and have not been so trained.

Using the same methods, 111,000 firefighters serving rural communities (27 percent of the total) are estimated to serve in fire departments where no certification of firefighters as Firefighter Level I or II has taken place.

Note that there may be other firefighters—possibly many other firefighters—who lack certification serving in departments where some firefighters are certified. These firefighters are not reflected in the 111,000 figure cited above. Conversely, some departments where no one is certified may be providing a local equivalent of certification.

Emergency medical services

For rural fire departments, protecting fewer than 2,500 population, 58 percent of fire departments now provide emergency medical services (EMS). An estimated 35 percent of personnel involved in providing EMS in these departments lack formal training. This percentage cannot be converted safely to estimates of numbers of personnel lacking formal training.

For departments that provide EMS to rural communities, this is the level to which some of their personnel have been certified (shown as percent of departments):

- no certification (1.3 percent);

- first responder (18.4 percent);

- basic life support (14.5 percent);

- first responder and basic life support (i.e., some to one level and some to the higher level) (35.0 percent);

- first responder, basic life support, and advanced life support (6.9 percent);

- basic life support, advanced life support, and paramedic (19.2 percent);

- first responder, advanced life support, and paramedic (3.1 percent); and

- advanced life support and paramedic (1.4 percent).

Since the four levels are progressive, with each level incorporating the skills and knowledge of the previous level, it is reasonable to assume that a combination answer (e.g., first responder and basic life support) indicates that some personnel in the department are certified to one of the levels, and other personnel are certified to another level.

By contrast, a department that responds with only one level presumably has all its certified personnel certified to that one level. In every case, it is possible that some assigned personnel are not certified to any level.

Hazardous material response

Roughly two-thirds of rural fire departments protecting fewer than 2,500 population (67.2 percent) now provide hazardous material response.

Requirements of the U.S. Environmental Protection Agency (EPA) and the U.S. Occupational Safety and Health Administration (OSHA) specify that all assigned personnel must have formal training. Only 14 percent of departments serving rural communities are compliant with these requirements. An estimated 50 percent of personnel involved in hazmat response in rural communities lack formal training. This percentage cannot be converted safely to estimates of numbers of personnel lacking training.

Certification levels were assessed using the same format as for EMS, with the following results for departments serving rural communities:

- no certification (7.2 percent);

- awareness (43.7 percent);

- operational (10.0 percent);

- technician (1.7 percent);

- awareness and operational (24.7 percent);

- awareness and technician (1.6 percent);

- operational and technician (1.2 percent); and

- awareness, operational, and technician (10.0 percent).

Since the three levels are progressive, with each level incorporating the skills and knowledge of the previous level, it is reasonable to assume that a combination answer (e.g., awareness and technician) indicates that some personnel are certified to one level and other personnel are certified to another level. By contrast, a department that responds with only one level presumably has all its certified personnel certified to that level. In every case, it is possible that some assigned personnel are not certified to any level.

Wildland firefighting

Roughly one-ninth (11.2 percent) of departments serving rural communities say they do not provide wildland firefighting. This is the lowest percentage for any community size.

In these departments, an estimated 45 percent of assigned personnel lack formal training. This percentage cannot be converted safely to estimates of numbers of personnel lacking training.

Technical rescue

For rural fire departments, protecting fewer than 2,500 population, 43.5 percent of fire departments now provide technical rescue. More than half (56 percent) of personnel performing this duty lacked formal training. This percentage cannot be converted safely to estimates of numbers of personnel lacking formal training.

Programs to maintain and protect firefighter health

An estimated 88 percent of departments serving rural communities do **not** have a program to maintain basic firefighter fitness and health, such as is required in NFPA 1500, *Standard on Fire Department Occupational Safety and Health Program*. An estimated 363,000 firefighters serve in departments without such programs. Roughly half (49 percent) of departments have programs for infectious disease control.

Fire Prevention and Code Enforcement

Some of the greatest value delivered by the U.S. fire services comes in activities that prevent fires and other emergencies from occurring or that moderate their severity when they do occur.

Here are the percentages of departments serving rural communities that provide each of selected programs:

- plans review (20.6 percent;)

- permit approval (10.4 percent);

- routine testing of active systems (11.3 percent);

- free distribution of home smoke alarms (19.9 percent);

- juvenile firesetter program (5.8 percent);

- school fire safety education program based on a national model curriculum (40.6 percent); and

- other prevention program (14.1 percent).

The program with the highest reported participation was school fire safety education programs based on a national model curriculum, where the majority of U.S. fire departments reported conducting such a program. This is one of the few programs in this section where there is some independent information regarding participation, and that information would suggest that implementation of a school-based fire safety curriculum following a national model exists is closer to 5 percent of fire departments rather than the reported 53 percent.

This large discrepancy may be a matter of interpretation. For example, many fire departments provide presentations to schools (e.g., puppet shows) in which the content is based on the content of some national model fire safety curriculum. Such presentations would qualify as a program of the sort asked about, but they would in practice have little educational value. Therefore, considerable caution should be shown when considering the reported practices for this particular program.

Here is who conducts fire-code inspections in rural communities (percent of departments):

- no one (39.4 percent);

- full-time fire department inspectors (3.3 percent);

- in-service firefighters (10.9 percent);

- building department (12.1 percent);

- separate inspection department (11.8 percent); and

- other (22.1 percent).

Here is who determines that a fire was deliberately set (percent of departments; multiple answers were possible):

- fire department arson investigator (14.5 percent);

- regional arson task force investigator (10.8 percent);

- State arson investigator (73.3 percent);

- Incident Commander (29.3 percent);

- police department (14.1 percent);

- contract investigator (1.3 percent);

- insurance investigator (15.1 percent); and

- other (9.7 percent).

Facilities, Apparatus, and Equipment

Fire stations

In rural communities, departments averaged 1.20 fire stations per department. This is the average that would result if, for example, 80 percent of departments have one station and 20 percent have two. Note that a fire station may have two or more firefighting companies, each attached to a particular apparatus, such as an engine/pumper. Here are the percentages of stations with specific potential needs:

- station over 40 years old (34.2 percent);

- station lacking backup power (71.7 percent; and

- station not equipped for exhaust control (91.7 percent).

In addition to needs associated with the condition of fire stations, there are also questions about needs with respect to the number and coverage of fire stations. The number and coverage needed are those required to achieve response with sufficient fire suppression flow within a target period of time. The information contained in the Needs Assessment Survey is not sufficient to perform such a calculation, but a simplified version is possible.

The *Fire Suppression Rating Schedule* of the Insurance Services Office includes a number of guidelines and formulas to use in performing a complete assessment of the adequacy of fire department resources, but for this simplified calculation on adequacy of number of fire stations, Item 560 has a basis: "The built-upon area of the city should have a first-due engine company within 1½ miles and a ladder-service company within 2½ miles."* For this simplified calculation, we can use these two numbers as a range for the maximum distance from any point in the community to the nearest fire station.

NFPA 1710 states its requirements in terms of time, specifically, a requirement that 90 percent of responses by the initial arriving company shall be within 4 minutes. If the first-response area is considered as a circle with the fire station in the middle, and if emergency calls are evenly distributed throughout the response area, then 90 percent of responses will be within 95 percent of the distance from the fire station to the boundary of the response area.**

If the average speed of fire apparatus is 21 mph, as it might be in the downtown area of a city, then the 4-minute requirement corresponds to a 1.5-mile requirement. If the average speed of fire apparatus is 36 mph, as it might be in a suburban or rural area, then the 4-minute requirement corresponds to a 2.5-mile requirement. In a very rural community, the average speed could be even higher, and the allowable distance would be even greater.

Note the limitations in this assumption: Item 560 implies that a larger maximum distance is acceptable for parts of the community that are not "built-upon"; this will be especially relevant for smaller communities. This larger maximum distance may or may not be on the order of the 2.5 miles cited for ladder-service companies responding in the built-upon area, so the use of 2.5 miles as an upper bound for calculation is done for convenience rather than through any compelling logic. Item 560 does not reflect variations in local travel speeds or the need for adequate fire flow by the responding apparatus; those issues are addressed elsewhere in the *Fire Suppression Rating Schedule*. This guideline is not a mandatory government requirement or a consensus voluntary standard.

To use this guideline with the data available from the Needs Assessment Survey, it is necessary to have a formula giving the maximum distance from fire station to any point in the community as a function of data collected in the survey. The Rand Institute

---

*Fire Suppression Rating Schedule. New York: Insurance Services Office, Inc., Aug. 1998, p. 28.

**If r is the distance from station to boundary, then the size of the response area is πr^2, and the radius of a circle with area equal to $0.9\pi r^2$ will be $r\sqrt{0.9}$ or approximately $0.95r$.

developed such a formula for expected (i.e., average) distance as part of its extensive research on fire deployment issues in the 1960s and 1970s.[†]

The formula has been developed and tested against actual travel-distance data from selected fire departments for both straight-line travel and the more relevant right-angle travel that characterizes the grid layout of many communities. It has been developed assuming either a random distribution of fire stations throughout the community or an optimal placement of stations to minimize travel distances and times.

The formula is called the square root law: Expected distance = $k \sqrt{(A/n)}$

where k is a proportionality constant

A is the community's area in square miles

n is the number of fire stations

Note the limitations of this approach, cited by the Rand authors: Most importantly, it ignores the effect of natural barriers, such as rivers or rail lines. It assumes an alarm is equally likely from any point in the community. It assumes a unit is always ready to respond from the nearest fire station.

If one further assumes that response areas can be approximated by circles with fire stations at the center, then expected distance equals one-half of maximum distance. If response areas are more irregularly shaped, expected distance will be a smaller fraction of maximum distance.

With these assumptions, the number of fire stations will be sufficient to provide acceptable coverage, defined as a maximum travel distance that is less than the ISO-based value, if the following is true:

$A - \frac{1}{2}(n)(D_{max})^2/(k^2) < 0$

where

A is the community's area in square miles

n is the number of fire stations

D_{max} is the maximum acceptable travel distance (1.5 miles or 2.5 miles)

k is the Rand proportionality constant, which is assumed to be for right-angle travel and is 0.6267 for random station location and 0.4714 for optimal station location

Estimates of need can be based on four calculations (i.e., two possible maximums for travel distance times two possible location protocols for fire stations). It may be appropriate to use the larger maximum distance for smaller communities. In fact, as noted, if the average speed achievable by fire apparatus is well above 36 mph, an even larger

---

[†]Warren E. Walker, Jan M. Chaiken, and Edward J. Ignall, eds. *Fire Department Deployment Analysis.* Publications in Operations Research series of the Operations Research Society of America, New York: Elsevier North Holland, 1979, pp. 180-184.

maximum distance is justified under NFPA 1710. Note also that NFPA 1720, the standard for volunteer fire departments, has no speed of response or distance requirement, reflecting the fact that very low population densities in the smallest communities mean the number of people exposed to long response times may be very small.

For rural communities, this results in an estimate that 73 percent of departments lack sufficient fire stations to achieve a maximum travel distance of 2.5 miles, assuming optimal station location, and 81 percent of departments lack sufficient stations, assuming random station location.

Remember the many limitations of this calculation procedure, however; a more complete calculation should be performed before drawing conclusions with regard to any particular community.

Apparatus

Rural communities have an average of 2.32 engines/pumpers in service per department and an average of 0.32 ambulances per department. (The latter number is what would result if one in three departments has one ambulance and the rest have none.) Here is the age distribution for the engines/pumpers in departments serving rural communities:

- 0 to 14 years old (35 percent);

- 15 to 19 years old (15 percent)—4,838 engines/pumpers;

- 20 to 29 years old (28 percent)—8,602 engines/pumpers; and

- 30 or more years old (22 percent)—6,854 engines/pumpers.

Vehicle age alone is not sufficient to confirm a need for replacement, but it is indicative of a potential need, which should be examined.

Only 4.3 percent of departments serving rural communities have ladder/aerial apparatus, but only 19.7 percent of those communities have any buildings that are at least four stories high.

Personal protective equipment and clothing

In departments protecting rural communities, an estimated 49 percent of emergency responders on a single shift are equipped with portable radios. It is further estimated that 64 to 72 percent of those radios are water-resistant and 69 to 84 percent are intrinsically safe in an explosive atmosphere. Also, 12 percent of departments have a number of reserve radios at least equal to 10 percent of the inservice radios.

The range indicates two approaches to the "Don't Know" responses. The higher numbers assume that the "Don't Know" responses are all cases of unrecognized need, because one would expect fire department managements to be aware if their radios have these sophisticated features. The lower numbers use the more conventional assumption that "Don't Know" respondents look like the other respondents, so the former are statistically allocated over the latter.

Also, in rural communities, an estimated 52 percent of emergency responders on a shift or otherwise on duty are equipped with self-contained breathing apparatus (SCBA). An estimated 53 percent of those SCBA units are at least 10 years old.

Also, in rural communities, an estimated 58 percent of emergency responders on a single shift are equipped with Personal Alert Safety System (PASS) devices.

The radio, SCBA, and PASS percentages cannot be converted safely to number of needed devices, because the need is for sufficient units to equip all personnel on a shift, not all personnel in the department. The number of personnel per shift is not known.

In rural communities, an estimated 10 percent of firefighters—or an estimated 42,000 firefighters—are **not** equipped with their own personal protective clothing. These unprotected rural firefighters represent roughly three-fourths of the estimated 57,000 total firefighters lacking such protective clothing in communities of all sizes.

It is further estimated that 45 percent of personal protective clothing in rural fire departments is at least 10 years old.

Fire Prevention Programs

Rural communities are much more likely to have significant gaps in code enforcement than are larger communities. No one conducts fire code inspections in 39 percent of rural communities (less than 2,500 population), compared to 26 percent of communities of 2,500 to 4,999 population, 15 percent of communities of 5,000 to 9,999 population, 6 percent of communities of 10,000 to 24,999 population, and 0 to 1 percent of all larger communities.

Fire departments do not conduct plans review in 79 percent of rural communities, compared to 66 percent of communities of 2,500 to 4,999 population, 27 percent of communities of 5,000 to 9,999 population, 52 percent of communities of 10,000 to 24,999 population, and, at most, 15 percent of communities of all larger population sizes.

Permit approval, routine testing of active systems, and other code-enforcement activities show similarly large gaps.

Of the 61 percent of rural communities where someone does conduct fire-code inspections, the largest share for providers was "Other" with 22 percent, compared to 3 percent for full-time fire department inspectors, 11 percent for in-service firefighters, 12 percent for building inspectors, and 12 percent for a separate inspection department. The "other" category could involve any or all of such arrangements as State agencies, regional authorities, or contract inspectors.

With regard to fire prevention programs or activities other than those related to code enforcement, 80 percent of rural fire departments have no program for free distribution of smoke alarms, compared with 20 to 30 percent of departments protecting communities of at least 100,000 population. Rural communities and larger cities have the highest percentage of need for such programs, based on inference from their higher rates of poverty.

Juvenile firesetter programs are offered by 6 percent of rural fire departments compared to well over a majority of fire departments serving communities of 25,000 or more population.

Two-fifths (41 percent) of rural fire departments offer school fire safety education programs based on a national model curriculum, compared with 60 to 70 percent of fire departments serving all larger communities combined.

Communications and Communications Equipment

An estimated 85 percent of fire departments serving rural communities can communicate by radio with their Federal, State or local partners at incident scenes. Interestingly, the percentage that can **declines** as the size of the department increases, the reverse of every other question so far. Of those who can communicate with their partners, 39 percent said they can communicate with all their partners, 44 percent with most of their partners, and 17 percent with only some of their partners.

Fifty-eight percent (58 percent) of rural departments said they have a map coordinate system with sufficient standardization of format to provide effective functionality in directing the movements of emergency response partners. Of the departments with such a system, 8 percent use a longitude/latitude coordinate system, 1 percent use the military grid, 1 percent use a State plan coordinate system, 86 percent use a local system, and 4 percent use an unspecified other system of map coordinates.

A local system is unlikely to be usable with global positioning systems (GPS) or familiar to, or easily used by, nonlocal emergency response partners, such as Urban Search and Rescue Teams, the National Guard, and State or national response forces. Moreover, interoperability of spatial-based information systems, equipment, and procedures will likely be rendered impossible beyond the local community under these circumstances.

Only 8 percent of fire departments serving rural communities do not have any 9-1-1 type telephone communication system for reporting emergencies. More than half (59 percent) have enhanced 9-1-1, another 32 percent have basic 9-1-1, and 1 percent have some other unspecified type of 3-digit system.

In rural communities, only 8 percent of fire departments have primary responsibility for dispatch operations. Roughly one-third (32 percent) have that responsibility lodged with the police department, another one-third (34 percent) with a combined public safety agency, 2 percent with a private company, and the remaining 25 percent with an unspecified other agency or entity.

The majority (53 percent) of fire departments serving rural communities have a backup dispatch facility. Forty-one percent (41 percent) have Internet access for the department. For those departments with Internet access, the majority (58 percent) have one access point at the one fire station. Another 7 percent have individual access for all personnel, 7 percent have access at headquarters but not at their multiple fire stations, 3 percent have one access point per station at multiple stations, and the remaining 25 percent have unspecified other access.

Ability to Handle Unusually Challenging Incidents

Fire departments were asked about their ability to handle four types of unusually severe and challenging incidents, only one of which involved a fire. These have to do with the increasingly important first responder role of fire departments.

In addition to asking whether such incidents were within the department's scope, the survey asked whether fire departments could handle such incidents with local personnel and equipment and whether a plan existed to support effective coordination with nonlocal resources and partners.

Technical rescue and EMS at structural collapse with 50 occupants

The first type of incident was a technical rescue with EMS at a structural collapse of a building with 50 occupants. Fifty-six percent (56 percent) of fire departments serving rural communities said such an incident is **not** within the scope of the department.

In rural communities, 10 percent of fire departments said they could handle such an incident with local trained personnel, 35 percent said they would need trained personnel from outside the local area, and, as noted, 56 percent said such an incident was outside their scope.

Switching to equipment, 9 percent said they could handle such an incident using local specialized equipment, while 35 percent said they would need to go outside the local area for needed equipment.

Thirteen percent of fire departments said they had a written agreement for using nonlocal resources, while 21 percent said they had a plan that was not written, and 11 percent said they had no plan for using nonlocal resources. Again, 56 percent said such an incident was outside their scope.

Hazmat and EMS for incident involving chemical/biological agents and 10 injuries

The second challenging incident was one involving hazmat and EMS for an incident involving chemical/ biological agents and 10 injuries. (Note that casualty counts of 100 to 1,000 are not unusual in planning for chemical/biological agent weapons of mass destruction.)

Fifty-three percent (53 percent) of fire departments serving rural communities said this type of incident was outside their scope.

Similarly to the structural collapse incident, 9 percent of departments said they could handle a chem/bio incident of the type specified with local trained personnel, while 37 percent said they would need nonlocal personnel; 8 percent of departments said they could handle such an incident with local specialized equipment, while 39 percent said they would need nonlocal equipment.

Finally, 13 percent of departments said they had a written agreement to guide use of nonlocal resources, while 24 percent said they had a plan that was not written, and 10 percent said they had no plan. As noted, 53 percent said such an incident was outside their scope.

Wildland/Urban interface fire affecting 500 acres

The third type of incident involved a wildland/urban interface fire affecting 500 acres. Only 28 percent of rural departments said this type of incident was outside their scope.

Thirty percent (30 percent) of departments said they could handle such an incident with local trained personnel, while 42 percent said they would need nonlocal personnel. Also, 26 percent of departments said they could handle such an incident with local specialized equipment, while 46 percent said they would nonlocal equipment.

Finally, 32 percent said they had a written agreement to guide the use of nonlocal resources, while 35 percent said they had a plan that was not written, and only 5 percent said they had no plan.

The U.S. Forest Service has made a considerable effort to create formal networks and plans to move resources from wherever they are to wherever they are needed. Their efforts appear to have borne fruit, because the current State of planning is also much better for these incidents than for the other types of incidents considered.

Mitigation of a developing major flood

The fourth and final type of incident was mitigation of a developing major flood. Sixty-one percent (61 percent) of rural fire departments said this type of incident was not within their scope. The study could not separate communities near a major waterway from communities that were not.

For personnel, 12 percent of rural fire departments said they could handle this type of incident with local trained personnel, while 27 percent said they would need nonlocal personnel. Also, 10 percent of departments said they could handle this type of incident with local specialized equipment, while 28 percent said they would need nonlocal personnel.

Only 8 percent said they had a written agreement to guide the use of nonlocal resources, while 20 percent said they had a plan that was not written, and 11 percent said they had no plan.

New and Emerging Technology

Only 8 percent of rural fire departments now own thermal imaging cameras, compared to 24 percent of all fire departments. Another 4 percent have plans to acquire one within a year, and 23 percent have plans to acquire one within 5 years, but 64 percent have no plans to acquire one.

Only 1 percent of rural fire departments have mobile data terminals, and most departments (91 percent) have no plans to acquire them.

Only 1 percent of rural fire departments have advanced personnel location equipment, and most departments (87 percent) have no plans to acquire them. The survey did not provide details on what constituted advanced personnel location equipment, which raises the possibility that departments differed in their views of the kind of equipment that would qualify as such.

Only 1 percent of rural fire departments have equipment to collect chemical or biological samples for remote analysis, and most departments (95 percent) have no plans to acquire them.

Survey Methodology

The Fire Service Needs Assessment Survey was conducted as a census, with appropriate adjustments for nonresponse. The choice of a census approach rather than a random sample approach was based on two considerations:

First, the survey is a specific requirement of PL 106-398 in Section 1701, Sec. 33(b), and the larger act is designed to provide the U.S. fire service with appropriate assistance for their legitimate needs. Given this intended application, there was general agreement that fire departments would view the survey as an opportunity rather than a burden, an opportunity that every department would wish to be given.

Second, current usage of some of the types of equipment and training to be addressed in the survey was believed to be sufficiently rare that the study would need the largest possible base for analysis.

The NFPA used its own list of local fire departments as the mailing list and sampling frame of all fire departments in the U.S.

The content of the survey was developed by NFPA, in collaboration with an ad hoc technical advisory group consisting of representatives of the full spectrum of national organizations and related disciplines associated with the management of fire and related hazards and risks in the U.S.

One-third (34 percent) of rural fire departments responded to the survey, and 60 percent of their surveys were received in time to be included in the analysis, which means roughly 2,700 rural fire departments were included.

Onduty Fatalities of Volunteer Firefighters

Each year since 1977, NFPA has studied onduty firefighter fatalities in the U.S. This section will report on the deaths of volunteer firefighters who were members of community fire departments. Because community size is not recorded in this casualty database, it is not possible to separate out the members of rural fire departments. NFPA statistics do indicate that nearly all rural firefighters are volunteers and roughly half of all volunteer firefighters serve in rural communities.

From 1995 through 2004, 597 volunteer firefighters died as the result of injuries or illnesses contracted while on duty. The number annually has ranged from a low of 49 in 1998 to a high of 70 in 1999.

Sudden cardiac death has been the number-one cause of firefighter fatalities among volunteers, accounting for 307 of the deaths (51 percent). Among the victims of sudden cardiac death, 90 were known to have suffered prior heart attacks or undergone bypass surgery or angioplasty/stent placement. Another 58 had severe arteriosclerotic heart disease, although they may not have been aware of it.

The next major cause of fatal injury was vehicle crashes (110 deaths or 18 percent). Among the 110 crash victims, 45 were driving or riding in personally owned vehicles (41 percent) and another 27 (25 percent) were driving or riding in water tenders. Unlike the career fire service, private vehicles are used commonly by firefighters to respond to fire stations or fire scenes, and water tenders are used commonly in rural areas where hydrant systems do not exist.

Firefighter fatalities are grouped by the assignments in which firefighters were engaged at the time of fatal injury or illness. The largest proportion were responding to or returning from alarms (218 deaths, or 37 percent). Victims in this category include 97 of the crash victims and 95 of the victims of sudden cardiac death.

The next largest proportion of deaths involved firefighters engaged in activities on the fireground (199 deaths, or 33 percent). Sudden cardiac death accounted for 113 of these fatalities.

In this 10-year period, the victims ranged in age from teenagers to 90 years old. Ninety of the victims were over the age of 65.

Appendix C.   *List of Experts Sharing Information for this Project*

The following were involved in detailed one-on-one conversations with project staff regarding the rural fire problem in America. Many of these discussions took place by telephone; however, some of them were face-to-face.

1. **Dicky Bain,** Fire Captain, former Fire Chief, Navajo Nation Fire Rescue

2. **Michael Clendenin**, Electrical Safety Foundation Institute (EFSI)

3. **Ken Brubaker**, National Rural Electric Cooperative

4. **Thom Danenhower**, Montana Department of Public Health and Human Resources, CDC smoke alarm project

5. **Steve Davidson**, Georgia Division of Public Health, CDC smoke alarm project

6. **Royal Edwards**, Technical Director, CSIA Technology Center, Chimney Safety Institute of America

7. **Craig Encinas**, AZ, President, Native American Fire Chiefs Association

8. **Pedro Flores**, former Fire Marshal, Bureau of Indian Affairs, Navaho Nation

9. **Jackie T. Gibbs**, Fire Chief, Marietta Fire Department, Marietta, GA, International Association of Fire Chiefs

10. **Christine Grammes**, National Rural Electric Cooperative

11. **Ernest Grant**, President, American Burn Association

12. **Tom Haynes**, Kentucky Injury Prevention and Research Center, CDC smoke alarm project

13. **Melissa Heeke**, Chimney Safety Institute of America

14. **Mark Jackson**, Centers for Disease Control and Prevention/National Center for Injury Prevention and Control/Division of Unintentional Injury Prevention smoke alarm project

15. **Mary Krom**, Alaska Program for the Prevention of Fire Related Injuries, CDC smoke alarm project

16. **Thomas Kuntz**, Fire Chief, Red Lodge Rural District #7, Red Lodge, MT, International Association of Fire Chiefs

17. **Mark W. Light**, CEM, Deputy Executive Director, International Association of Fire Chiefs

18. **Kay Lowder**, South Carolina State Department of Health and Environmental Control, CDC smoke alarm project

19. **Anne Mayberry**, former Executive Director, Electrical Safety Foundation Institute (EFSI)

20. **Robert H. McCool**, Program Manager, Community Injury Prevention Program, Kentucky Injury Prevention and Research Center

21. **Miriam McGaugh**, Oklahoma State Department of Health—Injury Prevention Services, CDC smoke alarm project

22. **Lori Moore**, Assistant to the General President for Technical Assistance and Information Resources, International Association of Firefighters

23. **Jimmy C. Parks**, MS, RN, Outreach Coordinator. Arkansas Children's Hospital Burn Center, Vice Chair, Burn Prevention Committee, American Burn Association

24. **John Paull**, Firefighter, Butte, MT, Fire Department, International Association of Fire Chiefs

25. **Toni Perkins**, Contractor with the Arkansas Department of Health, administering the Arkansas-CDC smoke alarm project

26. **Rio Grande Fire Chiefs** and officials serving Rural Areas, subgroup from National Association of Hispanic Firefighters meeting

27. **Gary C. Scott**, Fire Chief, Campbell County Fire Department, Gillette, WY, International Association of Fire Chiefs

28. **Heather Schafer**, Executive Director, National Volunteer Fire Council

29. **Philip C. Stittleburg**, Fire Chief, Lafarge, WI, Chairman of the Board, National Volunteer Fire Council

30. **Lisa Tenbrink Wolf**, Kansas Dept. of Health Education, CDC smoke alarm project

31. **Margaret Wilson**, Lexington, MS CDC smoke alarm project

32. **Gil Wood**, Results Champion, Office of Air Quality Planning and Standards, Program Implementation and Review Group, Environmental Protection Agency

33. **Courtney Yohe**, National Rural Health Association, Alexandria, VA.

As a followup, two group meetings took place at NFPA headquarters. Meeting #1 took place October 24 and 25, 2005, and included agencies and organizations that were stakeholders in rural communities and most likely had not been doing fire safety education programs. This group was invited to give suggestions for model programs that worked in rural areas and could be adapted to the rural fire problem. They were also a resource that could assist in the development of networks for effective delivery of relevant fire safety programs to rural areas. Some of the subject matter experts listed above also attended this meeting, and are listed here as well:

1. **Mark Jackson**, Unintentional Injury Prevention and Control, Centers for Disease Control and Prevention, CDC smoke alarm project

2. **Kay Lowder**, South Carolina State Department of Health and Environmental Control, CDC smoke alarm project

3. **Anne Mayberry**, former Executive Director, Electrical Safety Foundation Institute (EFSI)

4. **Jimmy C. Parks**, MS, RN, Outreach Coordinator. Arkansas Children's Hospital Burn Center, Vice Chair, Burn Prevention Committee, American Burn Association

5. **Mary Robertson Begay**, Hard Rock, AZ, Indian Health Service Injury Prevention Program on Navajo Nation

6. **Diane Schoenbauer**, Minnesota Valley Electric Cooperative, National Rural Electric Cooperative Association

7. **Tom Spurgeon**, Heartland Rural Electrical Cooperative, Inc., National Rural Electric Cooperative Association

8. **Tasha Toby**, Washington, DC, Safe Kids Worldwide

9. **Noah James West**, former Fire and Emergency Services Coordinator, Louisiana State University at Eunice

10. **Courtney Yohe**, National Rural Health Association, Alexandria, VA.

Meeting #2 took place October 27 and 28, 2005, and included representatives from the major fire service organizations and associations. In addition, groups that had been tasked with installation of smoke alarms and distribution of fire safety education were also in attendance. The intent of this meeting was to look at some of the unique challenges in rural communities and develop strategies to mitigate the fire problem in rural America. Some of the subject matter experts listed above also attended this meeting, and are listed again:

1. **Scott W. Adams**, Assistant Chief/Fire Marshal, Park City, UT, Immediate Past President, International Fire Marshals Association

2. **Clint Cobbins**, Tchula, MS, CDC Smoke Alarm project

3. **Steve Davidson**, Georgia Division of Public Health, CDC smoke alarm project

4. **Craig Encinas**, AZ, President, Native American Fire Chiefs Association

5. **Jackie T. Gibbs**, Fire Chief, Marietta Fire Department, Marietta, GA, International Association of Fire Chiefs

6. **Larry Holmberg**, Massachusetts Call/Volunteer Firefighters Association, National Volunteer Fire Council

7. **Mark Jackson**, Centers for Disease Control and Prevention/National Center for Injury Prevention and Control /Division of Unintentional Injury Prevention smoke alarm project

8. **Mark Larson**, Idaho State Fire Marshal, Chair—Fire Data and Information Committee, National Association of State Fire Marshals

9. **Stephen Sawyer**, NFPA liaison to the International Fire Marshals Association

10. **Gary C. Scott**, Fire Chief, Campbell County Fire Department, Gillette, WY, International Association of Fire Chiefs

11. **Margaret Wilson**, Lexington, MS, CDC smoke alarm project

Appendix D.   *Gap Analysis*

Rural communities suffer a fire death rate per million population that is more than twice the rate in small cities and large towns and considerably higher than even the fire death rates in the largest cities. For that reason, rural communities are the areas where it is most important and most urgent to target effective fire safety programs. The gap between that need and the current reality is the subject of this gap analysis.

This research project has developed information through a literature review and a series of informal discussions with individuals and groups, all on the following points:

- What is distinctive about the size and characteristics of the rural fire problem?

- What makes the rural fire problem particularly difficult to reduce?

- What are the advantages of working in a rural environment?

- What are key elements of successful fire safety programs designed for the rural environment?

What is the gap analysis?

The proposed gap analysis was intended to focus on this very generally defined type of gap:

> Gap between what is needed for success in rural fire safety programs and what is available in existing model and other programs.

Note that this gap is not limited to what is distinctive about the rural environment.

Types of fire safety programs

Fire safety professionals sometimes speak in terms of three "Es"—education, engineering, and enforcement. Safety can be improved by changes in behavior, directly achieved by education, or indirectly achieved by well-enforced codes and standards, which cause people to change their behavior. Safety can be improved by engineered changes to living environments, compelled by well-enforced codes and standards, or through voluntary action.

Gaps in code enforcement

Rural communities are much more likely to have significant gaps in code enforcement than are larger communities. No one conducts fire code inspections in 39 percent of rural communities (less than 2,500 population), compared to 26 percent of communities of 2,500 to 4,999 population, 15 percent of communities of 5,000 to 9,999 population, 6 percent of communities of 10,000 to 24,999 population, and 0 to 1 percent of all larger communities.

Fire departments do not conduct plans review in 79 percent of rural communities, compared to 66 percent of communities of 2,500 to 4,999 population, 27 percent of communities of 5,000 to 9,999 population, 52 percent of communities of 10,000 to 24,999 population, and at most 15 percent of communities of all larger population sizes.

Permit approval, routine testing of active systems, and other code-enforcement activities show similarly large gaps.

Of the 61 percent of rural communities where someone does conduct fire-code inspections, the largest share for providers was "Other" with 22 percent, compared to 3 percent for full-time fire department inspectors, 11 percent for inservice firefighters, 12 percent for building inspectors, and 12 percent for a separate inspection department. The "other" category could involve any or all of such arrangements as State agencies, regional authorities, or contract inspectors.

Only this limited data are available to demonstrate the size of the gap for rural communities in the area of code-enforcement fire prevention and fire safety activities, and none of our interviews or literature reviews identified any examples of successful efforts by individual rural communities to address this gap in innovative ways.

Public education and other voluntary fire safety programs

The rest of the gap analysis tends to focus on outreach programs where participation is voluntary.

An analysis follows in the form of a sequence of statements, with commentary on each, regarding what is challenging about fire safety programs in the rural environment, the available evidence on the magnitude and nature of the challenge, and the best ideas we have heard and seen on ways to address these challenges through successful fire safety programs.

Challenge: DISTANCE: Rural areas are areas with a low density of population per unit area (square mile, square kilometer). This significantly raises the cost of any door-to-door delivery of fire safety programs.

The U.S. Census Bureau defines "rural" as a community with less than 2,500 population. Other definitions use different thresholds and may include density criteria and/or criteria based on proximity to cities or other population concentrations. Many statistics are defined by metropolitan versus nonmetropolitan rather than nonrural versus rural. By either definition, rural communities account for roughly one-fifth of the U.S. population.

In recent years, the metropolitan statistical areas of the U.S. typically have had 80 percent of the U.S. population in 20 percent of the area. This means metropolitan areas have a population density 16 times the density in nonmetropolitan areas. That, in turn, means the average distance between two points is four times higher in a nonmetropolitan area than in a metropolitan area. In addition, rural areas are the least dense areas in nonmetropolitan areas. Therefore, the distance ratio for rural versus nonrural may be even higher.

In a door-to-door program, quadrupling distance means more time and more cost per household. Consequently, it may result in driving from place to place rather than walking from place to place.

Complication: Rural areas have lower rates of usage of and access to most forms of mass media.

The perception is that rural areas have fewer television and radio channels, fewer newspapers, less coverage by cable systems, and less access to the Internet.

Myth: Rural areas have less Internet access than urban areas.

A 2003 survey by the U.S. Department of Commerce (DOC) found that 54.1 percent of rural households had Internet access (61.9 percent had computers) compared to 54.8 percent of urban households (61.7 percent had computers) and 49.3 percent of central city households (56.9 percent had computers). [*A Nation Online*; *http://www.ntia.doc. gov/ntiahome/dn/index.html*] There is no significant difference between rural and urban access for Internet access.

This myth underscores the need to verify key facts. Widespread perceptions often lag the reality in the field or are formed from impressions shaped by unrepresentative experience.

Sometimes, a statement that is a myth in its commonly stated form is close to another relevant statement that is true. The same DOC survey found that among the poorest households (under $5,000 per year), the poor rural households had markedly less Internet access (20.0 percent) than the poor urban households (28.4 percent) and even than the poor central city households (24.3 percent).

Another related statement that is true is that rural fire departments are far less likely to have Internet access than are fire departments in larger communities. Only 41 percent of rural communities had Internet access for fire departments, compared to at least 63 percent for fire departments in any larger sized community, and at least 93 percent for communities of 25,000 population or more.

Therefore, it is not true that rural areas generally lag other areas with respect to Internet access, but it is true for the poorest rural households and for the fire departments that serve rural communities. Therefore, it is true for the highest-risk rural population—the one most in need of fire safety programs—and it is true for the agencies that are the most natural choices to lead enhanced delivery of fire safety programs in their communities.

Comparable measures of rural versus nonrural access to other forms of mass media have proven elusive. It seems clear, however, that rural choices are often fewer, such as fewer radio and television stations within broadcast range and fewer and smaller newspapers to cover local events. In other words, mass media are less well-established in rural communities, which makes mass media less effective and less attractive for use in fire safety programs. Under these conditions, door-to-door distribution is the more effective option.

Challenge: TRUST: There is a consensus perception that rural households are slower to give trust to people.

Similarity and familiarity are considered key characteristics in gaining trust. Similarity may involve ethnicity, religion, age, culture, region, and language. Familiarity means a lower threshold for trust if someone unknown is known to a third party who is known and trusted.

It is not clear whether this challenge is distinctive to rural areas. Some very successful fire safety programs in major cities have encountered trust issues in their target populations and solved them by working with and through faith networks.

It was suggested in some of the project discussions that urban dwellers are more accustomed to dealing with strangers, suggesting a higher degree of mobility and turnover in such communities. A U.S. Census Bureau analysis of movement from 1995 to 2000 found that 59 percent of nonmetropolitan dwellers were in the same residence in both years, compared to 53 percent of metropolitan dwellers (and 49 percent of central city dwellers). The survey also found that 80 percent of nonmetropolitan dwellers were in the same county in both years, compared to 79 percent of metropolitan dwellers (and 79 percent of central city dwellers). [Table 1, *Migration and Geographic Mobility in Metropolitan and Nonmetropolitan America*: 1995 to 2000, *http://www.census.gov/prod/2003pubs/censr-9.pdf*] This supports the notion of a somewhat less mobile rural population, but the difference is not overwhelming, and it is virtually nonexistent at the same-county level.

Another version of this point would be that the lower density of rural areas means a rural dweller is surrounded with far fewer people within his/her range of potential contact and interest. Under these circumstances, it is almost inevitable that urban dwellers will have a much higher fraction of their encounters with people they do not know outside specified roles such as merchant and customer. This would lead back to the same conclusion, namely that urban dwellers do not have the option of dealing only with people they know well and, therefore, trust. Rural dwellers do have the option, and the perception is that many of them use it.

There is anecdotal evidence that many people who grew up in rural areas and later went to "the big city" as adults are returning to the rural communities of their youth, possibly to tend to aging parents or to recapture the familiar and comfortable feelings of their youth. These returnees would be more experienced in dealing with strangers and might be seen as strangers themselves. Furthermore, any vital rural community will have some growth through in-migration, and some of those "transplants" will bring perceptions and habits from larger communities. At the same time, however, there continues to be anecdotal evidence of rural communities perceived as so lacking in opportunities that every young person who can leave does so.

With so many thousand rural communities, it would not be surprising if all these phenomena are true in some rural places. It has not been possible to identify hard evidence of just how much truth there still is in the generic, insular, rural community with sharply distinguished insiders and outsiders.

To the extent that there is some truth to these characterizations, there are implications for fire safety programs. A program without a trusted local advocate is unlikely to succeed. Once the basis of trust has been established with some, however, there may be a very rapid spread of trust to the entire community. There is a lower "tipping point" in a rural community.

Solution: EXISTING NETWORKS: By building a fire safety program around existing networks, both the trust and the distance problems can be addressed.

A network, as the term is used here, consists of a central organization with existing relationships for particular purposes with a larger group of people in the community. Several existing networks common to rural areas have been identified as potentially valuable to and supportive of fire safety programs. The central organizations for these identified existing networks are as follows:

- fire departments (nearly all volunteer in rural communities);

- health care (including both public health personnel and the individual private care providers, who may be the only ones located in a rural community);

- churches and other faith groups;

- schools;

- Fire Corps;

- area agencies on aging, senior citizen centers, meals-on-wheels and other older adult organizations (people age 65 and older are a high-risk group for fire death);

- rural electrical cooperatives;

- national safety organizations, such as the american red cross and safe kids worldwide; and

- cooperative extension programs.

Farm-related groups were considered, but farm dwellers constitute only 1 percent of the U.S. population and only 5 percent of the rural population. Farm-related groups have only a fraction of the reach such groups had a century ago.

All of these networks have gaps in coverage. For example, fire departments are unlikely to have established relationships with their communities if their only activity is fire suppression. Some rural communities do not even have a fire department to call their own but rather obtain fire suppression services from a neighboring or regional authority.

A 2001 fire service needs assessment study found that 20 percent of rural fire departments had programs of free distribution of smoke alarms, compared to 40 percent of nonrural fire departments. Also, 41 percent of rural fire departments had school fire safety education programs, compared to 66 percent of nonrural fire departments. And only 11 percent of rural fire departments had inservice fire code inspections by firefighters, compared to 25 percent of nonrural fire departments. Other potential outreach programs showed similar large gaps between rural and nonrural departments.

These statistics suggest that rural fire departments will need to be recruited to service as leaders or participants in expanded fire safety programs for their communities and that, if they agree, they will have to build up networks and relationships from a less advanced position than typically exists in larger communities.

The federally funded **Fire Corps** program, launched in December 2004, organizes community volunteers to help fire departments by performing nonoperational or nonemergency roles, such as fundraising, and public education activities, including education in local schools, home safety checks, and smoke alarm installation programs. Fire Corps is administered by the National Volunteer Fire Council (NVFC) with assistance from the IAFC. Because Fire Corps is part of a national network, there is support for the local fire corps to get involved in fire safety education activities in local areas.

Area Agencies on Aging (AAAs) are in communities all across the country. They plan, coordinate, and offer services that help older adults remain in their homes aided by services such as Meals-on-Wheels, homemaker assistance, and whatever else it may take to make independent living a viable option. AAAs also work through local senior centers. Older adult organizations provide access to a portion of the 12 percent of the national population (or the somewhat higher percentage of the rural population) who are in the high-risk upper age groups.

While it is not known whether those who frequent senior centers include proportional participation by all ethnic, religious, education, and income groups, one must consider the possibility that participants will be drawn disproportionately from the lower-risk part of the age group. Fire service personnel will have to work with AAAs and other older adults organizations to make sure that their programs not only are distributed through senior center but also are reaching people in their own homes.

Schools provide access to the 5- to 17-year-old age group (grades K to 12), which constitutes 18 percent of the U.S. residential population, but this age group has below-average risk except for the very youngest children. To be fully effective, school-based programs will probably need to provide access to other family members and/or provide significant effects for life. Neither of these larger effects has been well documented for even the most effective school programs.

Rural electrical cooperatives have a history that gives them potentially greater interest in the quality of life of their customers than one might expect from a typical power company, and that history may also give them a higher degree of existing trust. Those advantages of trust would presumably be greatest for programs targeted on electrical equipment fires, because rural electrical cooperatives interact with their customers primarily on matters involving electrical equipment. Ordinarily, the targeting of electrical equipment fires would be hard to justify, because electrical distribution equipment ranks only sixth in share of fatal home fires, well behind smoking materials, intentional, heating equipment, and cooking equipment. However, in rural areas, the share of fatal fires involving electrical distribution equipment is higher than it is for nonrural areas. Therefore, this targeting **would** make sense within an overall strategy of targeting the larger parts of the rural fire problem.

Health care and faith networks probably have the broadest coverage of any of the existing networks. Like any network, these networks may vary from one to another in the degree to which their leaders and members see their missions broadly and strategically versus narrowly and traditionally. Individuals with a broad vision are more likely to be open to the value of participation in an ambitious fire safety program. Individuals with a narrow vision still may be willing to permit use of their resources in a more passive and less demanding fashion.

The USDA Services has networks through universities which have been granted federally owned land. These are called "land-grant" universities. They have mandated outreach responsibilities and they provide educational programs on a variety of subjects. A volunteer group that often works with them is the Extension Homemakers. The Extension Homemakers often get involved in safety education. There are extension offices in most counties throughout the United States.

Other national volunteer and service organizations such as the American Red Cross and Safe Kids Worldwide may have local or regional chapters that also provide outreach to small towns. Their outreach will vary throughout the country.

Because most of the existing community networks do not have a pre-existing interest in or commitment to fire safety program, the networks will have to be recruited to the cause.

The ideal situation is for a far-reaching and effective network to agree to full partnership and coleadership of a fire safety program. In other words, at least one network is needed that will agree to do the "heavy lifting" required for a program to be a success. At least one network must be found to agree to participate at this level.

Some networks may be unable or unwilling to commit to this level of participation but may be willing to actively recruit their members and constituents to the program. For example, a church might include in the weekly sermon a strong endorsement of the program. As another example, a health care facility might arrange for its doctors or other staff to encourage everyone they see in any capacity to participate in the fire safety program.

Finally, some networks may be willing to participate only in a passive manner. Imagine a church that allows third parties to place a program invitation on its bulletin board or in its newsletter.

Any help is welcome, of course, but a certain minimum level of help is necessary if a program is to be successful. It is possible that a local network is part of a national structure, and it may be that the local network can be approached effectively by way of their national counterparts. Many churches have a national hierarchy or organization, and health care facilities work with many national government agencies and nonprofit associations.

Finally, every community will have individuals and organizations that are not connected to a network but which are potential sources for funds, other resources, and/or volunteers.

Advantage: INFLUENTIAL FEW. The perception is that rural communities have a simpler, smaller group of key influential people. It is literally possible to ask who the most influential person is in town—the one best able to get things done—and receive a specific answer that proves accurate when tested.

This means that when trying to set up a program by recruiting one or more networks and persuading those networks to take a substantial, demanding role, there may be only a few people who need to be sold on the proposition. In a smaller group, there is less bureaucracy and greater flexibility. The advantage is that there are fewer steps to success under these conditions. The risk, however, is that there may be no viable Plan B if the influential few are unconvinced.

In the long run, a smaller network can make a program more fragile, because the program's success will be tied so closely to the interest, energy, and continued involvement of a couple of people. Any disruption in that leadership core will be difficult to overcome. Therefore, it is essential to seek ways to broaden support and participation for a program early and often and to seek to institutionalize the program constantly in ways that will insulate it from variations in the fates and personal arcs of individuals.

During the early stages of a program, it is especially important that program managers take steps to anticipate and plan for setbacks and other "bad news." The ability to respond quickly and effectively to any shock to the system is likely to be tested, and the need should be anticipated.

A common theme to all these tasks and stages in successful programs is the need for persistence, even doggedness. It takes persistence to enlist the influential few. It takes persistence to attract sponsors and partners. It takes persistence to gain access to the target population, even with pre-existing trust. Sustained effort, continuity, and follow-up are all phrases that came up repeatedly and with emphasis in our conversations with our experts.

Challenge: VOLUNTEERS. Any program run by volunteers is subject to challenges familiar to anyone involved in a nonprofit organization.

Recruiting and retaining sufficient numbers of workers, supervisors, and leaders is an ongoing challenge to any program. This is complicated by a societal trend of declining participation in many kinds of volunteer activities. Training volunteers for effectiveness is another challenge.

Some aspects were given special emphasis by our experts in discussions of rural fire safety programs.

Turnover was described as a major reason for program failure. Growing time pressures in the modern life and a tendency for rural dwellers to work somewhere outside their community both were cited as societal trends that give people less room to volunteer. Awkward and ineffective leadership succession arrangements—such as electing a new volunteer fire chief every year—were cited as problems.

Many, if not most, of these challenges are common to nonprofit organizations, and there is now a considerable literature on the management of such organizations. The use of rewards and recognitions, fundraising techniques, and management structures and procedures to address these challenges are all among the issues discussed at length in this literature.

Challenge: LOCAL RELEVANCE: Part of recruiting workers and constituents to a program is making it locally relevant, as judged by the community.

Local relevance is a matter of perception. It is achieved or increased by gathering information on local preferences for program design, priorities, and the like, and then modifying the programs to better reflect those preferences.

Gathering the information can be done in a number of ways. The most assured but expensive approach is to ask the people in the community systematically. Less expensive versions of this approach include surveying a sample of the community, which still can be expensive, and interviewing a small group of opinion leaders, which can be misleading if those leaders do not know their constituents' opinions well. Important sub-populations may be illiterate or speak only a language other than English, or both; this adds to the complexity and cost of gathering their opinions. All the familiarity and similarity issues that impact trust and complicate program delivery also complicate any attempt to ask what people do and don't want.

This challenge puts a premium on workers who have a detailed understanding of the constituents, what they like and don't like, what they need and want, and how to relate to them. The earlier-cited "influential few" are likely to know the opinions of the people of their communities.

Compared to systematically asking people, passively observing and listening is less assured of success. This is a less costly, more opportunity-driven approach that is also less intrusive and, as a result, less likely to create resistance among constituents. The downside is that this approach can be dominated by people with strong opinions, loudly expressed, and those people and their opinions may not be representative of the community. An advantage of this approach is the use of observation, which means the information is not limited to constituent perceptions but is calibrated to some degree by observable facts. A program whose messengers know to look for fire hazards outside the original scope of the project is a program with a built-in procedure to grow the program for steadily greater value and impact in successive years.

The approach least assured of success is acting on pre-existing information regarding local preferences. The danger here is that this information may be a collection of myths, stereotypes and prejudices having only occasional correspondence with the community realities. These pre-existing beliefs also tend to be more general and more sweeping than the typically more complex reality, as the earlier-cited myth about Internet access illustrates.

Another complication in attempts to increase local relevance is that a community may not be the best judge of what it needs, from the point of view of objectively recognizing primary factors in creating their heightened fire risk. Rural dwellers are not immune from stereotyping of their neighbors and, like people everywhere, they form their perceptions of what is most or least dangerous from a flow of information that is subject to a variety of distortions.

This puts a premium on people who are good listeners and creative customizers but also thoroughly understand the programs and the reasons for their specific features and priorities. It also puts a premium on model programs that are designed for easy replication in different settings and designed for easy but careful and sound customization.

Notwithstanding these reservations about the accuracy and credibility of pre-existing beliefs, our experts offered a number of characterizations of rural dwellers, which led to some ideas about how to incorporate these characterizations into customization of model programs and approaches:

- **Heating equipment** accounts for a much larger share of fatal fires in rural than in nonrural communities. The share for **electrical distribution equipment** is also higher in rural communities, and the share for intentional fires (arson) is typically lower. Strategies for reducing equipment fires include changes in equipment choices and behavioral changes in usage, maintenance, and the like.

Heating equipment fires are usually space heater fires, where space heaters include both portable heaters and fixed space heaters, such as wood stoves and fireplaces with inserts. The most effective fire safety program would probably be to switch everyone from space heating to central heating, where risks are much lower regardless of power or fuel choice. This, however, would be prohibitively costly, not only initially, but in terms of ongoing costs of heating. Educational programs to prevent heating equipment fires are long established and widely used, and they should not change much, if at all, in a rural setting. Only the relative emphasis on particular types of equipment might change.

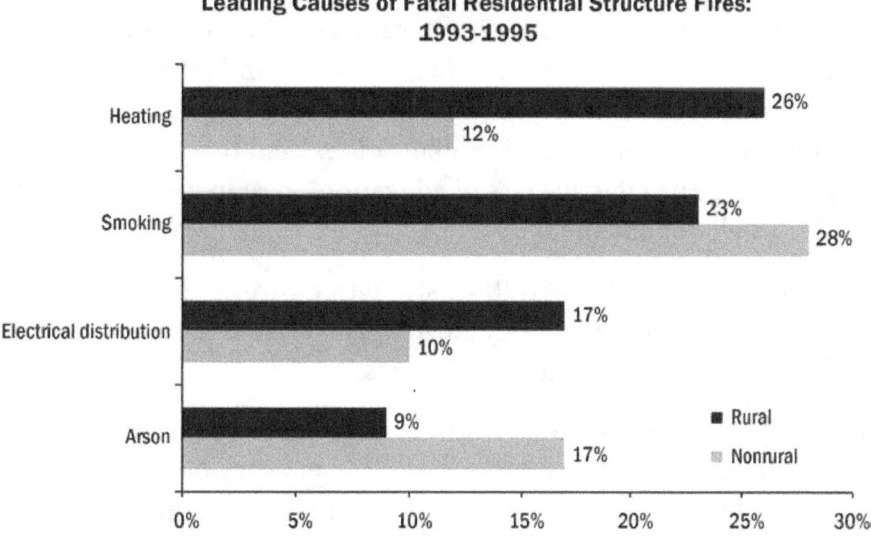

Leading Causes of Fatal Residential Structure Fires: 1993-1995

Source: USFA's *The Rural Fire Problem in the United States.*

• **Electrical distribution equipment fires** are led by wiring system fires and cord/plug fires, most of which involve extension cords. There are a number of behavioral strategies applicable to the latter, but there is little or no field evidence or research to suggest how much leverage one can achieve on this fire problem by providing education that is not complemented by changing the equipment people have.

One of the important gaps is better evaluative information on effective strategies directed at electrical distribution fires.

• **Equipment fires of all types.** There are a number of national government programs that provide grants or low-interest loans to rural residents (often limited to low-income households) to upgrade housing equipment for safety reasons.

The EPA has a Wood Stove Changeout Campaign (*http://www.epa.gov/wood-stoves/changeout.html*) which provides financial incentives to replace older wood stoves with newer heating equipment, wood-burning or other, that pollutes less. Such a change could lead to reduced fire risk as well.

The Rural Housing Service within the USDA has a program of grants and low-interest loans for low-income rural households. Upgrading of heating equipment is one of the purposes that could be eligible, and other repair projects that remove safety hazards appear eligible as well. The Rural Housing Service also has a program to assist rural communities in building capability for fire departments and other emergency responder agencies.

Rural electrical cooperatives have indicated a willingness to provide safety information to their customers and constituents not only on electrical system equipment and electric-powered appliances but on all types of household equipment.

Although second to heating-equipment fires, **smoking-material fires** are still among the leading causes of fatal fires in rural communities. Risks can be reduced by changing the heat source, the first fuel, or behavior. Mattresses and upholstered furniture sold in the U.S. have been subject to cigarette-resistance requirements or programs for almost 40 years, but items older than that are worth identification and replacement. The effectiveness of education on reducing risks of cigarette fires has not been well documented.

Initiatives to require reduced-ignition-strength cigarettes have succeeded in three States (New York, Vermont, and California) and are being pursued in many others. None of this is specific to rural communities.

Within the outdoor fire problem, rural areas have their primary problem with **outdoor burning**, while urban areas have their primary problem with intentional fires. Outdoor burning may seem like a purely behavioral problem, but the severity of an outdoor fire, in terms of loss of life and property, can depend on environmental specifics that affect whether, where, and how fast fire will spread. There are a number of programs available, which make particular sense for rural communities.

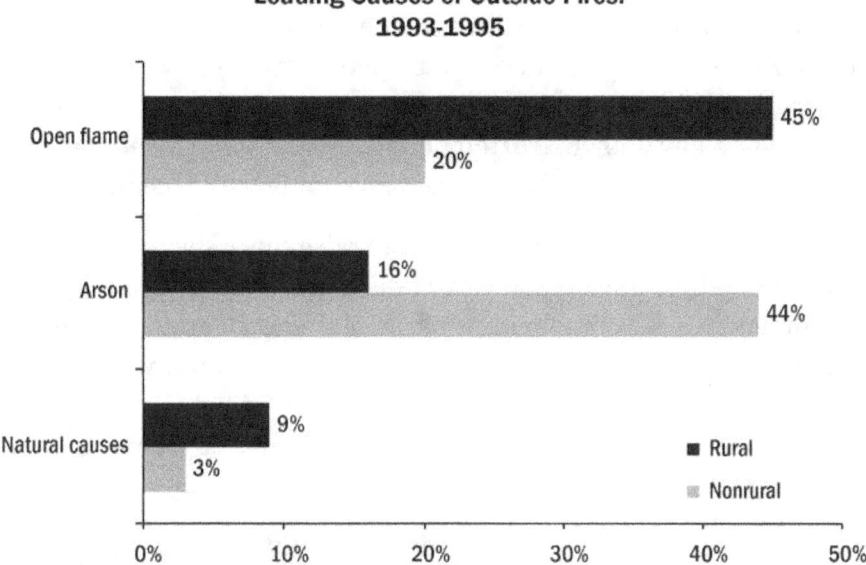

Leading Causes of Outside Fires: 1993-1995

Source: USFA's *The Rural Fire Problem in the United States*

- **Water supplies** can be a problem in rural areas. For firefighters, the absence of an onsite water supply suitable for firefighting, such as from hydrants connected to a piped public water supply under pressure, has several negative effects. One of these is to add a significant task of finding and accessing a local source for fire-fighting water. That task can require time and personnel who would otherwise be performing other tasks, it can delay the time when water is first put on the fire, and it can complicate the incident command task for the officer in charge.

  Sprinkler systems in homes also should be considered. Fire hoses, on average, use more than 8-1/2 times the water that residential sprinklers do to contain a fire. In remote areas, self-contained water tanks can be used to supply residential fire sprinkler systems. Therefore, when considering improvements to water supplies, consider options that will provide an infrastructure for more use of residential fire sprinklers as well as supporting firefighter hose operations.

- According to our experts, rural dwellers are more likely to believe in **self-reliance**, but possibly also more likely to believe in fate. The former includes a resistance to receiving charity, but this is paired with a greater willingness to give help. Those who accept a program may be more likely than their nonrural counterparts to take the next step of volunteering to work in the program. Most importantly, the program needs to position its actions as something other than charity and needs to have sincere, persuasive, predetermined responses to give anyone who reacts to the program as charity.

- Rural dwellers tend to be more **conservative politically**. This may play out as resistance to far-away authorities, including national or State government and including national model standards and codes. At the same time, there may be a greater comfort with local authorities, who are people they know. Programs that seem to originate elsewhere and be led by people from elsewhere may attract automatic push-back. This is a variation of the trust issue. To be accepted, a program needs to be vetted by known, trusted people. By the same token, programs with such vetting may be more likely to attract near-universal acceptance.

Training for the volunteer workers also may raise this issue, depending on who the volunteers are. If they are already knowledgeable about fire safety and its intricacies, volunteers may welcome training from a credible national or regional body. If not, they may resist the use of national standards of training as unsuited to the needs and ways of the local community. Volunteer firefighters, for example, can be resistant to standards that are perceived as originating from the interests and concerns of career firefighters alone.

- Rural communities operate more on **word-of-mouth**. Programs need to look for ways to make this informal communication work for them. For example, if the first participants (constituents) in a program have a positive experience and identify known, respected, key people in the community who are supporting the program, they can pass this on in advance of the workers and help to presell the program, reducing the need to establish trust from scratch at every household.

- Rural communities have fewer distractions and fewer things going on. The threshold for an activity to be seen as a welcome **change of pace** is considered lower in rural communities. If the leaders make the programs interesting and exciting, they may give the targeted constituents extra reasons to participate and to do so enthusiastically, without necessarily adding anything to the cost.

- Rural communities are regarded as more traditional, more attuned to **ritual and taboo**, than nonrural communities. The challenge is to make that orientation work for a program of behavioral change, by making the change seem to be **not** change but rather a more complete fulfillment of community values—even more traditional than the current traditions. This is a considerable challenge in framing the program to fit with local values, and if it is approached as an exercise in public relations or spin, constituents will recognize the lack of sincerity and candor. This will create a backlash that will be hard for the program to survive. But if the program's leaders are from the community, they are likely to have worked out in their own minds the reasons why the program's goals harmonize with traditional values of the community.

- Make the programs **simple but sound**. Any simplification that does not reduce program soundness or effectiveness improves the chances of program acceptance and successful implementation. This is not limited to rural communities.

- Even something as simple as local terminology (phraseology) can play a critical role in positioning a worker as a familiar friend or an alien outsider.

Challenge: RURAL-SCALE ORGANIZATION: Rural programs face challenges of scale, complicated by a need to minimize costs and decentralize control, that are greater than those faced in nonrural communities.

If it is true that a successful rural fire safety program needs to (1) operate primarily door-to-door so that communication will be trusted and effective, (2) customize program materials for local relevance without harming effectiveness, (3) use messengers from the community so they are familiar to the community, and (4) train messengers to national standards to achieve program effectiveness, then rural programs will face higher costs for administration, training, recruitment, travel, and possibly even program materials than their nonrural counterparts.

The cost aspects can be addressed by partnering and wide-ranging efforts to attract donations. The nonprofit management guides extensively discuss these matters.

A separate issue is the organizational and management challenge of organizing and running programs on this scale. If every few hundred people, representing a distinct community, needs their own customization of materials and their own pool of volunteers, then one is likely to need a separate program coordinator for every community, and multiple people to support the customization exercises and try to keep all versions of the program as harmonized and effective as possible, consistent with the need to customize in order to gain entry. This army of volunteers needs an army of supervisors, and the implications in terms of multiple levels of management are daunting.

The analogy, and precedent, is to national nonprofit organizations with active, large chapter structures. These national nonprofit organizations also struggle with variations in commitment, resources, and energy from one to another. They also struggle with the tension between centralized quality and consistency, control versus local autonomy, and the benefits that flexibility provides.

What is possibly unique about the rural environment is the very small size of the typical unit. Whereas a chapter might cover a State or a metropolitan area, each with hundreds of thousands or millions of people, the basic unit in a rural environment is a community of hundreds or at most thousands of people. The difference in scale from an organizational point of view is significant, and it is not clear that there is strong guidance available on how to operate effectively in this less populous environment. This is an important gap to be addressed.

Challenge: REGIONAL DIFFERENCES: The relative size and detailed characteristics of the rural fire problem differ, depending on the region of the country.

The U.S. Census Bureau divides the country into four primary regions: Northeast, north central, South, and West. (See Figure 1.)

The South region has by far the largest share of the U.S. population (36 percent in 2000, according to the 2006 *Statistical Abstract of the United States*), and its rural share (26 percent in 2000) is larger than the rural share of any other region. The South accounted for 46 percent of the total rural population in the U.S. in 2000.

In the past 10 years (1995 to 2004), the South's rural fire death rate was the highest of all the regions in 3 years, and the second highest in 5 years, making it the region with the largest number of years (8) in the top two. However, in the past 6 years, the South has had the highest regional rural fire death rate only once, which indicates that its rural fire death risk is now very like the other regions, after many years when it was clearly higher. From 2000 to 2004, the South narrowly had the highest rural fire death rate (29.0 deaths per million population) of all the regions.

Figure 1. Major Regions in the U.S.

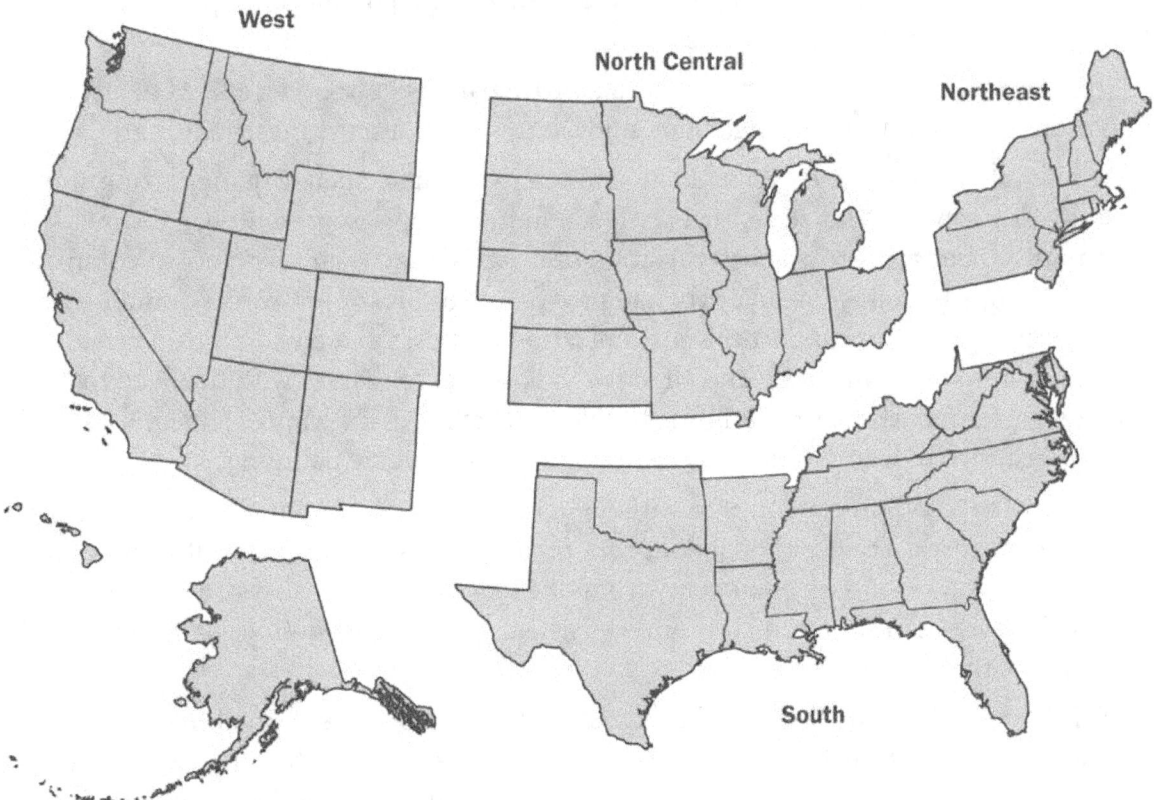

The rural South has nearly all of America's rural African-Americans, and nearly all of America's poor rural African-Americans. In March 1997, 90 percent of nonmetropolitan African Americans were located in the South, according to Table 3 in a 1998 Census Bureau study. [*http://www.census.gov/population/socdemo/race/black/tabs97/ tab03.txt*] In 1996, according to Table 15, 92 percent of nonmetropolitan African-Americans living below the poverty line were located in the South. [same Web site but tab15 rather than tab03]

African-Americans accounted for 14 percent of African-Americans and whites combined (excluding Hispanic whites) in metropolitan areas and 9 percent in nonmetropolitan areas, in March 1997 nationwide. The corresponding combined national African-American percentage in March 1997 was 13 percent. The African-American percentage was 20 percent in metropolitan areas versus 19 percent in nonmetropolitan areas in the South, compared to 10 percent in metropolitan areas versus 2 percent in nonmetropolitan areas in all other regions combined.

In 1996, African-Americans accounted for 15 percent of African-Americans and whites combined (excluding Hispanic whites), and the national percentage of African-Americans living below the poverty line was 37 percent. In 1996 in the South, African-Americans constituted 52 percent of poor African-Americans and whites combined (excluding Hispanic whites) in metropolitan areas versus 42 percent in nonmetropolitan areas. In the same period using the same inclusion criteria (African-Americans and whites only and excluding Hispanic whites), the African-American percentages for the other three regions combined were 35 percent of the metropolitan poor and 4 percent of the nonmetropolitan poor.

In 1996 for African-Americans and whites combined (excluding Hispanic whites), the percent of the population who were poor was 8 percent for all metropolitan areas combined versus 15 percent for all nonmetropolitan areas combined; 12 percent for the metropolitan South, versus 18 percent for the nonmetropolitan South; and 10 percent for the metropolitan part of all other regions combined, versus 12 percent for the non-metropolitan part of all other regions combined.

The implications of these statistics are that rural areas tend to be poorer than non-rural areas, but that difference is much more pronounced in the South. Rural populations tend to have a lower African-American share than do nonrural populations, but that difference is almost nonexistent in the South. The rural poor tend to have a lower African-American share than do the nonrural poor, but that difference is much more pronounced outside the South. In the rural South, when targeting the high-risk poverty population, white and African-American each account for nearly half that target population. In rural areas outside the South, almost none of the high-risk poverty population is African-American (though because of the limitations of these statistics, there may be significant shares of Hispanic whites, Native Americans, or Asians in the poverty population outside the South).

The heightened share of rural fire deaths involving heating equipment is as much a Southern phenomenon as it is a rural phenomenon. For example, as noted, in 1993 to 1995, heating equipment accounted for 26 percent of residential structure fire deaths in rural areas versus 12 percent in nonrural areas, while smoking materials accounted for 23 percent of residential structure fire deaths in rural areas versus 28 percent in nonrural areas. In 1993 to 1997, heating equipment accounted for 14 percent of residential structure fire deaths in the Nation as a whole (rural and nonrural) compared to 23 percent for smoking materials. In the South (rural and nonrural) in 1993 to 1997, heating had a 19 percent share, the same as smoking materials. In the West (rural and nonrural), heating had a 10 percent share, compared to 22 percent for smoking materials. In the north central (rural and nonrural), heating had an 11 percent share, compared to 24 percent for smoking materials. In the Northeast (rural and nonrural), heating had a 10 percent share, compared to 32 percent for smoking materials. [Regional statistics taken from Michael

J. Karter, Jr., *U.S. Fire Experience by Region*, Table 12, NFPA Fire Analysis & Research Division, January 2001.]

Of the four regions, the South has the most consistently mild and short heating season. Therefore, poorer households in the South are the ones who find it most feasible to try to use space heating exclusively, resulting in the fire experience repeatedly documented for space heating as compared with central heating.

In 2003, usage of central heating (warm-air furnace, steam, or hot water system) was highest in the north central region (90 percent) and Northeast (88 percent), where the entire region is subject to severe winters, while it was lower in the West (68 percent), where part of the region has consistently milder winters and part does not. Usage of central heating was lowest in the South (61 percent), where the entire region has consistently milder winters than the rest of the country.

Some of these facts are puzzling. The West is closer to the South in heating equipment usage (central versus space), but the West's heating fire problem looks more like the Northeast and north central regions, much lower than the South's. In addition, the leading type of space heating equipment in the South by far is electric heat pumps. Heat pumps, however, do not stand out as a specific type of equipment resulting in fires. These figures do not isolate rural dwellers or poor households, and it is possible the patterns are quite different for those high-risk groups.

Type of Primary Heating Equipment	Northeast	North Central	South	West
Warm-air furnace	40 %	81 %	59 %	65 %
Electric heat pump	1 %	2 %	24 %	6 %
Steam or hot water system	48 %	9 %	2 %	3 %
Floor, wall or pipeless furnace	2 %	2 %	4 %	13 %
Built-in electric units	6 %	4 %	2 %	7 %
Room heaters with flue	1 %	1 %	2 %	1 %
Room heaters without flue	0 %	0 %	4 %	0 %
Stoves	1 %	1 %	1 %	2 %
Fireplaces	0 %	0 %	0 %	0 %
Cooking stoves	0 %	0 %	0 %	0 %
None	0 %	0 %	0 %	1 %
Portable electric heaters	0 %	0 %	1 %	1 %
Other	0 %	0 %	0 %	0 %
Total	**100 %**	**100 %**	**100 %**	**100 %**

Source: 2004-2005 *Statistical Abstract of the United States*, Table 947.

Poor housing quality generally is a problem in the rural South. Three-quarters of the substandard housing units in the 1980s were in the South. In 1995, the 9 percent of the Nation's housing units in the nonmetropolitan South accounted for 21 percent of U.S. occupied units with moderate physical problems, 11 percent with severe problems, and 12 percent of the households with income below the poverty level. That last figure is consistent with the fact that the percentage of the population living in poverty, not limited to rural areas, is highest in the South (16 percent versus 13 percent in the West, 12 percent in the north central, and 11 percent in the Northeast, all in 1990 from the 1994 *Statistical Abstract of the United States*).

The South also has the highest proportion of housing units in manufactured homes—12 percent versus 3 to 7 percent in the other regions (from the 2004-2005 *Statistical Abstract of the United States*). More than half (56 percent) of all the manufactured homes in the country are in the South. This has historically been a factor in the elevated fire death rate in the South, because until very recently manufactured homes have had a higher fire death rate than conventional "stick-built" homes or apartments. However, now that most manufactured homes in use were built after the advent of the construction requirements of the U.S. Department of Housing and Urban Development (HUD), introduced in 1976, manufactured homes are no longer a high-risk environment. This may be part of the reason why fire death rates in the South are no longer consistently much higher than rates in other regions.

The north central region has the second largest share of the U.S. population (24 percent) and the second highest percentage of regional population in rural areas (28 percent). This means the north central has the second largest share of total rural population (28 percent). In the 1980s, rural areas were not consistently the community size with the highest fire death rates in the north central region.

In the past 10 years (1995 to 2004), the north central's rural fire death rate was the highest of all the regions in 1 year and the second highest in 4 years, making it the region with the lowest number of years as the highest region, but the second highest number of years (5) in the top two. In the past 6 years, the north central has had the highest regional rural fire death rate only once, and in 2000 to 2004 combined, the north central had the lowest rural fire death rate (22.8 deaths per million population) of all the regions.

The Northeast region has the fourth largest share of the U.S. population (20 percent) but the third highest percentage of regional population in rural areas (21 percent). This means the Northeast has the third largest share of total rural population (17 percent). In the 1990s, rural areas were often not the community size with the highest fire death rates in the Northeast.

In the past 10 years (1995 to 2004), the Northeast's rural fire death rate was the highest of all the regions in 2 years and the second highest in 1 year, making it the region with the lowest number of years (3) in the top two. In the past 6 years, the Northeast has had the highest regional rural fire death rate only once, and in 2000 to 2004 combined, the Northeast had the second lowest rural fire death rate (27.0 deaths per million population) of all the regions.

The West region has the third largest share of the U.S. population (21 percent) but by far the lowest percentage of regional population in rural areas (14 percent), which is why the West has by far the lowest share of total rural population (12 percent). However, these figures were as of 1990, and the West had by far the largest increase in nonmetropolitan population between 1990 and 2000 (21 percent compared to 12 percent in the South, 6 percent in the north central and 5 percent in the Northeast). [*Population Profile of the United States: 2000*, Figure 2-1, from *www.census.gov*] The West's share of total rural population probably has increased substantially, principally at the expense of the Northeast and north central regions.

In the past 10 years (1995 to 2004), the West's rural fire death rate was the highest of all the regions in 4 years but never the second highest, making it the region with the highest number of years as top region but the second lowest number of years (4) in the top two. In the past 6 years, the West has had the highest regional rural fire death three times, and in 2000 to 2004 combined, the West had the second highest rural fire death rate (28.2 deaths per million population) of all the regions.

The West has by far the highest Hispanic percentage in its population (24 percent in 2000 versus 12 percent in the South, 10 percent in the Northeast, and 5 percent in the north central, from Betsy Guzman, *The Hispanic Population: Census 2000 Brief*, C2KBR/01-3, May 2001, at *www.census.gov*). The Hispanic population nationwide has the highest percentage of its people employed in farming, fishing, and forestry (2.7 percent versus 0.5 percent for non-Hispanic white, 0.4 percent for African-American, 0.3 percent for Asian, and 1.3 percent for Native American, from Peter Fronczek and Patricia Johnson, Occupations 2000: Census 2000 Brief, C2KBR-25, August 2003, at *www.census.gov*). These statistics seem to come as close as any readily available to substantiating and quantifying the importance of the Hispanic farming population, including migrant workers, in the rural West.

The West has by far the highest percentage of Native Americans in its population (2.0 percent versus 0.4 percent in the Northeast, 0.6 percent in the north central, and 0.8 percent in the South). Native Americans are a high-risk group collectively and especially for those living on reservations. The West has by far the highest percentage of Native Americans on reservations in its population (0.7 percent versus 0.0 percent in the Northeast, 0.1 percent in the north central, and 0.4 percent in the South). Therefore, any rural programs for the West need to consider the distinct character of Native Americans, and reservations in particular, in their design and execution. Differences among different Nations (e.g., Navajo, Cherokee) also need to be considered as factors in customizing programs and delivery approaches. This applies to code enforcement programs as well; for example, the Navajo Nation does not have a fire code.

Related to the higher risk on reservations, Native Americans have the highest differential in fire death rate of any ethnic group when rural and nonrural areas are compared. For Native Americans, the fire death rate in rural areas is 2.6 times the rate in nonrural areas, compared to 1.6 for African-Americans and 1.3 for whites.

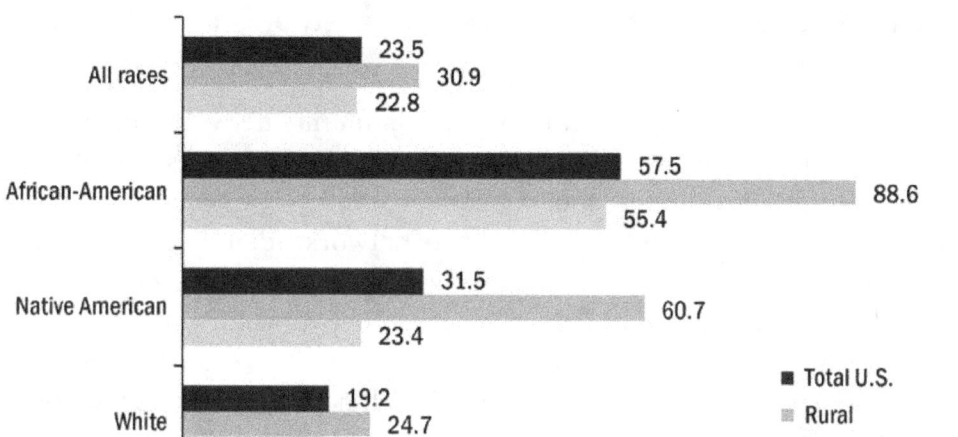

1983-1988 Fire Death Rates by Race

Source: USFA's *The Rural Fire Problem in the United States.*

The West has the second highest percentage of housing units that are manufactured homes (7 percent compared to 12 percent in the South, 5 percent in the north central, and 3 percent in the Northeast). As noted above, this may have been a factor increasing their rural fire death rate in the past, but is probably not such a factor now.

The West has the highest percentage of fires and property damage due to fire involving outdoor properties (55 percent in 1995 to 1999 versus 51 percent for the South, 48 percent for the north central, and 47 percent for the Northeast). Outdoor fires in rural areas are likely to be unintentional, involving open flame sources—a description that suggests open burning—while outdoor fires in nonrural areas are much more likely to be intentionally set.

Summary of gap analysis

- **Code enforcement versus voluntary programs:** Significant gaps in activity exist for both code-enforcement and voluntary fire safety programs. None of the research approaches identified innovative approaches or guidance for improved code enforcement in rural communities.

- **Distance:** The average distance between two households is at least four times greater in rural communities than in nonrural communities. Rural households may have less access to mass media than nonrural households, although the available hard data do not show a large difference. Door-to-door program delivery will cost more because of distance but may be more necessary if mass media is less effective. There is a gap in affordable program distribution methods for such an environment.

- **Trust:** While hard evidence is difficult to come by, there is a strong consensus that gaining trust can be difficult for an outsider in a rural community. There is a presumed pre-existing gap in trust, which any fire safety program must plan and act to address.

- **Existing networks:** Tapping into existing community networks appears to be the best way to achieve the necessary trust. Such existing networks also offer the possibility of sufficient volunteer foot soldiers to perform the necessary door-to-door delivery without inordinate cost. Candidate networks identified are fire service (almost entirely volunteer and often involved only in emergency response activities), religious, health care, and the specialized networks for certain target groups of senior centers, schools, and electric cooperatives. There are gaps in terms of verified guidance in how best to work with such networks for this kind of purpose, but the existing literature on management of nonprofit organizations is a start.

- **Influential few:** If existing networks are the preferred way to reach a widely dispersed population with serious trust issues, then working through the influential few in a community is the preferred way to achieve partnership with those networks and to identify other desirable or necessary modifications to program or delivery style to achieve best results in a particular community. There is a gap in specific guidance on how to approach this highly political step in a successful program's timeline.

- **Volunteers:** Major tasks include recruiting volunteers, retaining volunteers, training volunteers, and supervising and coordinating volunteers. The literature on nonprofit organization management can help. There is a gap in documented success stories focused on this part of successful programs.

- **Local relevance:** Model programs must be customized to reflect local needs, priorities and preferences. At the same time, however, every effort must be made not to compromise the effectiveness of the program or to adopt local stereotypes and myths as if they were facts. There is a gap in proven techniques to achieve this delicate balancing act, but there may be more guidance available for borrowing in the way that successful national nonprofit organizations operate with their chapters.

- **Rural scale organization:** A typical rural community, consisting of hundreds or at most thousands of people, is a much smaller base organizational unit than is commonly used in existing national ground-up programs. There is a major gap in proven, affordable, effective organizational techniques appropriate to this scale.

- **Regional Differences:** The rural South is distinguished by a much higher heating equipment fire problem associated with space heating and the lowest usage of central heating of any region. The rural South also is distinguished by very high poverty rates (though not so high as in the nonrural South) and a high African-American share (short of half) of that poverty population. The South has nearly half the Nation's rural population, which means it has disproportionate influence on national characterizations of the rural population, and the South tends to have the highest regional rural fire death rates, although this has been less true in recent years.

The rural West is distinguished by a larger share of Hispanic whites, including a significant number engaged in agriculture, and a larger share of Native Americans, including a larger share of those living on reservations. The West has a larger proportion of its fires outdoors, and rural areas have a much larger proportion of their outdoor fires involving open burning. The rural West is the fastest growing rural population by a wide margin.

The rural north central region and rural Northeast have characteristics more like each other than like either of the other two regions. They are much whiter and slightly poorer than their nonrural counterparts in the two regions. It is not clear whether these two regions have the higher shares for heating equipment and electrical distribution equipment that characterize the rural areas generally, because cause profiles have not been developed separately for rural versus nonrural areas in these two regions. However, the two regions have older housing, on average, than the South and West (in 2003, the median year of construction for occupied housing was 1956 in the Northeast, 1965 in the north central, 1974 in the West, and 1976 in the South), befitting the fact that national population growth has tended to concentrate further south and further west. Age of housing is known to correlate with higher rates of electrical distribution equipment fires. By contrast, these two regions have the highest rates of usage of central heating, which would suggest that they would not be expected to have heightened rates of heating equipment fires.

Appendix E. *Best Practices*

Fire safety professionals sometimes speak in terms of three "Es"—education, engineering, and enforcement. Safety can be improved by changes in behavior, directly achieved by effective education programs, or induced by well-enforced codes and standards. Safety can be improved by engineered changes to living environments, induced by well-enforced codes and standards, or through voluntary action.

This report collects guidance, in the form of recommendations, on best program practices that were developed from the rural project research.

Recommendation 1: Pay particular attention to the program's ability to reach all or nearly all of its target audience. Rural communities have special challenges in program distribution.

Example 1.1: Programs that operate through local codes and regulations need to address gaps in code enforcement, which tend to be greater in rural communities.

Rural communities are much more likely to have significant gaps in code enforcement than larger communities. Fire code inspections are not conducted in 39 percent of rural communities (less than 2,500 population), compared to 26 percent of communities of 2,500 to 4,999 population, 15 percent of communities of 5,000 to 9,999 population, 6 percent of communities of 10,000 to 24,999 population, and 0 to 1 percent of all larger communities.

Plans review, permit approval, routine testing of active systems, and other code-enforcement activities show similarly large gaps.

Example 1.2: Programs that are delivered in person need to address the lower densities and greater place-to-place distances in rural communities.

The average distance between two points is four times higher in a nonmetropolitan area than in a metropolitan area. In addition, rural areas are the least dense areas in nonmetropolitan areas. Therefore, the distance ratio for rural versus nonrural may be even higher. In a door-to-door program, quadrupling distance means more time and more cost per household. Consequently, it may result in driving from place to place rather than walking from place to place.

Example 1.3: Programs that are delivered via mass media need to be realistic about usage and access rates in rural communities for the selected mass media and the target population for the programs.

Rural communities tend to have fewer television and radio stations within broadcast range. Satellite television may bring far more stations within range, but satellite television, like cable television, may be less affordable for the low-income rural residents who tend to be the primary targets for fire safety programs.

As of a 2003 survey by the U.S. Department of Commerce, rural households had closed the access gap with respect to the Internet, but the gap was still large between rural poor and urban poor households.

Therefore, when planning for program delivery, it is important to consider the reach of selected mass media to target populations, which will tend to be the poorer members of the rural community, and it is important to obtain up-to-date data on the reach of those media. Access has been increasing rapidly, and some perceptions of rural needs have been overtaken by events.

Recommendation 2: Build a relationship of trust between those delivering a program and those targeted to receive the program.

It is not clear whether the establishment of trust, an essential condition for program success, is more difficult in rural areas, as some believe. Some very successful fire safety programs in major cities have encountered trust issues in their target populations and solved them by working with and through faith networks.

Available data do not clearly support the perception of a higher degree of mobility and turnover in nonrural communities. The influx of strangers seems to be roughly as common in communities of all sizes.

Regardless of whether or not the establishment of trust is a major challenge elsewhere, it is viewed as a challenge in rural communities. A program without a trusted local advocate is unlikely to succeed.

Once the basis of trust has been established with some, however, there may be a very rapid spread of trust to the entire community if the perception is correct that there is a lower "tipping point" in a rural community.

Recommendation 3: Build around existing networks.

A network, as the term is used here, consists of a central organization with existing relationships for particular purposes with a larger group of people in the community. Several existing networks common to rural areas have been identified as potentially valuable to and supportive of fire safety programs. The central organizations for these identified existing networks are as follows:

- fire departments (nearly all volunteer in rural communities);

- health care (including both public health personnel and the individual private care providers, who may be the only ones located in a rural community);

- churches and other faith groups;

- schools;

- Fire Corps;

- area agencies on aging, senior citizen centers, Meals-on-Wheels, and other older adult organizations (people age 65 and older are a high-risk group for fire death);

- rural electrical cooperatives;

- national safety organizations, such as the American Red Cross and Safe Kids Worldwide; and

- cooperative extension programs.

Farm-related groups were considered, but farm dwellers constitute only 1 percent of the U.S. population and only 5 percent of the rural population. Farm-related groups have only a fraction of the reach such groups had a century ago.

It is important to remember, however, that all of these networks have gaps in coverage. For example, fire departments are unlikely to have established relationships with their communities if their only activity is fire suppression. Some rural communities do not even have a fire department to call their own but instead obtain fire suppression services from a neighboring or regional authority. Rural fire departments are much less likely to conduct any of the major fire-loss prevention programs associated with fire departments, such as free distribution of smoke alarms (20 percent of rural departments versus 40 percent of nonrural departments). Therefore, rural fire departments are less likely to have an existing active fire safety network.

The federally funded Fire Corps program, launched in December 2004, organizes community volunteers to help fire departments by performing nonoperational or non-emergency roles, such as fundraising, and public education activities, including education in local schools, home safety checks, and smoke alarm installation programs. Fire Corps is administered by the NVFC, with assistance from the IAFC. Because Fire Corps is part of a national network, there is support for the local Fire Corps to get involved in fire safety education activities in local areas.

Area agencies on aging (AAAs) are in communities all across the country. They plan, coordinate, and offer services that help older adults remain in their homes, aided by services such as Meals-on-Wheels, homemaker assistance and whatever else it may take to make independent living a viable option. AAAs also work through local senior centers. Older adult organizations provide access to a portion of the 12 percent of the national population (or the somewhat higher percentage of the rural population) who are in the high-risk upper age groups.

While it is not known whether those who frequent senior centers include proportional participation by all ethnic, religious, education, and income groups, one must consider the possibility that participants will be drawn disproportionately from the lower-risk part of the age group. Fire service personnel will have to work with AAAs and other older adults organizations to make sure that their programs not only are distributed through senior center but also are reaching people in their own homes.

Schools provide access to the 5- to 17-year-old age group (grades K to 12), which constitutes 18 percent of the U.S. residential population. This age group, however, has below-average risk except for the very youngest children. To be fully effective, school-based programs will most likely need to provide access to other family members and/or provide changes in the safety knowledge and behaviors of schoolchildren that last throughout their lives. Neither of these wider or longer lasting impacts has been well documented for even the most effective school programs.

Rural electrical cooperatives emerged during the socially conscious New Deal period as a central element in bringing the full range of modern conveniences to previously isolated rural communities. With this historical context of a larger mission than simply delivering electricity to a new set of customers, these cooperatives may be seen by their management and their customers as concerned about the health and well-being of their community, not just about being paid on time. Such attitudes would make rural electrical cooperatives a potential network for rural fire safety programs, particularly programs targeted on electrical equipment fires. It is known that such equipment has a higher share of fatal fires in rural areas than in nonrural areas.

Health care and faith networks probably have the broadest coverage of any of the existing networks. Like any network, however, these networks may vary from one to another in the degree to which their leaders and members see their missions broadly and strategically versus narrowly and traditionally. Individuals with a broad vision are more likely to be open to the value of participation in an ambitious fire safety program. Individuals with a narrow vision still may be willing to permit use of their resources in a more passive and less demanding fashion.

The USDA Services has networks through universities which have been granted federally owned land. These are called "land-grant" universities. They have mandated outreach responsibilities and they provide educational programs on a variety of subjects. A volunteer group that often works with them is the Extension Homemakers. The Extension Homemakers often get involved in safety education. There are extension offices in most counties throughout the United States.

Other national volunteer and service organizations such as the American Red Cross and Safe Kids Worldwide may have local or regional chapters that also provide outreach to small towns. Their outreach will vary throughout the country.

Because most of the existing community networks do not have a pre-existing interest in or commitment to fire safety program, the networks will have to be recruited to the cause.

Any help is welcome, of course, but a certain minimum level of help is necessary if a program is to be successful. The ideal situation is for a far-reaching and effective network to agree to full partnership and coleadership of a fire safety program. In other words, at least one network is needed that will agree to do the "heavy lifting" required for a program to be a success.

Other networks may be unable or unwilling to commit to this level of participation but may be willing to recruit their members and constituents actively to the program. For example, a church might include in the weekly sermon a strong endorsement of the program. As another example, a health care facility might arrange for its doctors or other staff to encourage everyone they see in any capacity to participate in the fire safety program.

Finally, some networks may be willing to participate only in a passive manner. For example, a church might allow third parties to place a program invitation on its bulletin board or in its newsletter.

It is possible that a local network is part of a national structure, and it may be that the local network can be effectively approached by way of their national counterparts. Many churches have a national hierarchy or organization, and health care facilities work with many national government agencies and nonprofit associations.

Finally, every community will have individuals and organizations that are not connected to a network but which are potential sources for funds, other resources, and/or volunteers.

Recommendation 4: Try to work with a rural community's key influential people.

It is possible to ask who the most influential person is in town—the one best able to get things done—and receive a specific answer that proves accurate when tested. In other words, when trying to set up a program by recruiting one or more networks and persuading those networks to take a substantial, demanding role, only a few people may need to be sold on the proposition. In a smaller group, there is less bureaucracy and greater flexibility and, as a result, there are fewer steps to success. The risk, however, is that there may be no viable Plan B if the influential few are unconvinced.

In addition, a smaller network can make a program more fragile, because the program's success will be tied so closely to the interest, energy, and continued involvement of a couple of people. Any disruption in that leadership core will be difficult to overcome. Therefore, it is essential to seek ways to broaden support and participation for a program early and often and to seek to institutionalize the program constantly in ways that will insulate it from variations in the fates and personal arcs of individuals.

Recommendation 5: Plan for delays and setbacks, and be ready to adapt or respond as needed

During the early stages of a program, it is especially important that program managers take steps to anticipate and plan for setbacks and other "bad news." The ability to respond quickly and effectively to any shock to the system is likely to be tested, and the need should be anticipated.

A common theme in successful programs is the need for persistence, even doggedness. It takes persistence to enlist the influential few. It takes persistence to attract sponsors and partners. It takes persistence to gain access to the target population, even with pre-existing trust. Sustained effort, continuity, and followup are all phrases that came up repeatedly and with emphasis in conversations with experts.

Recommendation 6: Consult available books and other resources on best practices in the management of nonprofit organizations, with particular attention to the oversight and effective use of volunteers.

The recruiting, training, and retention of sufficient numbers of workers, supervisors, and leaders are ongoing challenges to any program. Turnover is a major threat to program success.

Societal trends that complicate these challenges further include declining participation in many kinds of volunteer activities, growing time pressures in modern life, and a tendency for rural dwellers to work somewhere outside their community. In addition, awkward and ineffective leadership succession arrangements—such as electing a new volunteer fire chief every year—can be problems.

The use of rewards and recognitions, fundraising techniques, and management structures and procedures to address these challenges are all among the issues discussed at length in the literature on management of nonprofit organizations.

Recommendation 7: Modify model programs to reflect local conditions and provide local relevance.

Make sure, however, that the information used to identify local conditions is current and fact-based, and that the changes do not damage the program elements that are essential to its design and effectiveness.

Gathering the information can be done in a number of ways. The most assured but expensive approach is to ask the people in the community systematically. Less expensive versions of this approach include surveying a sample of the community, which still can be expensive, and interviewing a small group of opinion leaders, which can be misleading if those leaders do not know their constituents' opinions well.

Regardless of how information is gathered, important subpopulations may be illiterate or speak only a language other than English, which will add to the complexity and cost of gathering their opinions. In addition, all the familiarity and similarity issues that affect trust and complicate program delivery also complicate any attempt to ask what people do and don't want.

Compared to asking people systematically, listening passively is a less costly, more opportunity-driven approach that also is less intrusive and, as a result, less likely to create resistance among constituents. The downside is that this approach can be dominated by people with strong opinions, loudly expressed, and those people and their opinions may not be representative of the community.

Whether opinions are obtained passively by listening or actively by asking, there is a value to the use of observation. The facts obtained through observation can be used to calibrate the accuracy of the inherently more subjective expressed opinions. A program whose messengers are also observers can build a database on a wide range of hazards, not just those that are the focus of the current program, and this can provide direction to future expansion of the scope of the program, which can mean more safety and more impact for the community.

The approach least assured of success is reliance on pre-existing information regarding local preferences. The danger of this approach is that such information may be a collection of myths, stereotypes, and prejudices having only occasional correspondence with the community realities. These pre-existing beliefs also tend to be more general and more sweeping than the typically more complex reality, as the earlier-cited myth about Internet access illustrates.

Another complication in attempts to increase local relevance is that a community may not be the best judge of what it needs. Rural dwellers are not immune from stereotyping of their neighbors and, like people everywhere, they form their perceptions of what is most or least dangerous from a flow of information that is subject to a variety of distortions. Objective analysis of the correlation between fire experience and possible explanatory conditions or factors may show that the community's sense of what does and does not threaten it is not consistent with the facts.

This recommendation of customizing programs to local conditions puts a premium on the earlier cited "influential few" and any other workers who have a detailed understanding of the constituents, what they like and don't like, what they need and want, and how to relate to them. The recommendation also places a premium on people who are good listeners and creative customizers but also thoroughly understand the programs and the reasons for their specific features and priorities. It also puts a premium on model programs that are designed for easy replication in different settings and designed for easy but careful and sound customization.

The following are some common differences between the leading fire challenges in rural communities and those in other communities:

- Heating equipment accounts for a much larger share of fatal fires in rural communities than in nonrural communities. The share for electrical distribution equipment is also higher in rural communities, and the share for intentional fires (arson) typically is lower.

- Rural residents may be eligible for certain national government programs that provide grants or low-interest loans to rural residents (often limited to low-income households) to upgrade housing equipment for safety reasons. Some examples:

  - The EPA has a Wood Stove Changeout Campaign (*http://www.epa.gov/wood-stoves/changeout.html*) which provides financial incentives to replace older wood stoves with newer heating equipment, wood-burning or other, that pollutes less. Such a change could lead to reduced fire risk as well.

  - The Rural Housing Service within the USDA has a program of grants and low-interest loans for low-income rural households for several related purposes. Upgrading of heating equipment is one of the purposes that could be eligible, and other repair projects that remove safety hazards appear eligible as well. The Rural Housing Service also has a program to assist rural fire departments and other emergency responder agencies to expand their ability to serve the needs of their communities.

— Rural electrical cooperatives have indicated a willingness to provide safety information to their customers and constituents, not only on electrical system equipment and electric-powered appliances, but on all types of household equipment.

• Within the outdoor fire problem, rural areas have their primary problem with outdoor burning, while urban areas have their primary problem with intentional fires.

• Water supplies can be a problem in rural areas. This affects not only the effectiveness of fire suppression forces when fire occurs but also affects the cost and feasibility of fire sprinkler systems. For firefighters, the absence of an onsite water supply suitable for firefighting, such as from hydrants connected to a piped public water supply under pressure, has several negative effects. One of these is to add a significant task of finding and accessing a local source for firefighting water. That task can require time and personnel who would otherwise be performing other tasks, it can delay the time when water is first put on the fire, and it can complicate the incident command task for the officer in charge.

Recommendation 8: Consult the literature on rural safety and health program design and delivery.

The project literature review identified the following two articles, both by C. Neil Bull and Shari DeCroix Bane, that had extensive guidance directly on point for any summary of best program practices:

The first article, by C. Neil Bull and Shari DeCroix Bane. "Program Development and Innovation," appeared in *Journal of Applied Gerontology*, vol. 20, no. 2, June 2001, pp. 184-194, and reached these conclusions:

• The term "rural" describes a variety of settings, including "frontier" counties with populations less than six per square mile and rural counties that abut urban ones. Differences also exist between farm and nonfarm rural populations.

• Four core issues affect rural programs: geographic isolation, economic deprivation, human service infrastructure, and economies of scale. The distance that must be traveled is a part of the geographic isolation. Public transportation is close to nonexistent. Older adults may feel that giving up driving is not an option due to a lack of alternatives. Terrain and weather also can make driving difficult. Costs for long-distance calls, fuel, and travel time add up quickly.

• Economic deprivation is exacerbated by the tendency for many rural areas to rely heavily on one industry, activity, or service for local livelihoods. Economic shifts or plant closures can be devastating to residents' income and local tax revenues. Also, some Federal and State programs mistakenly assume that services can be provided at less cost in rural areas, and do not fund adequately. Finally, because rural incomes are lower and fewer foundations are rural, there are fewer charitable resources available for programs or for the matching funds necessary to qualify for some grants.

- The human service infrastructure has experienced consolidations and closings. There is a shortage of technical equipment and skilled personnel. Rural youth often move away, and more women are working, reducing the volunteer pool that might partially alleviate the lack of paid workers.

- Economies of scale are difficult because the numbers of people and suppliers are simply not there. In some cases, there is a sole supplier. Competitive bidding may not be possible.

- In addition to these four core issues, rural housing is said to be less well maintained, rural residents less educated, and nonfarm rural residents tend to be in poorer health. Moreover, the independence associated with rural life, particularly among the elderly, often results in a resistance to using or accepting services or assistance. Seven points are made about transferring urban programs to rural areas:

  1. Expectations may need to be scaled, back, particularly if success is defined as number of people served.

  2. It is often necessary to scale services to offer only the highest priority (as defined by the community), rather than offering the full range.

  3. Program duplication should be avoided, and offerings coordinated so that each agency offers programs it can do best.

  4. Rules and regulations should be handled with some flexibility as bookkeepers and accountants tend to be in short supply. Budget waivers should be sought when expenses will be higher than expected for items like long-distance calls and mileage.

  5. Do not expect economies of scale or more than one provider bidding.

  6. Create partnerships or reciprocal agreements so that the jurisdictional or administrative boundaries do not interfere with services.

  7. Plan for challenges in recruiting and keeping qualified personnel. Hire people with multiple competencies rather than narrow specialists.

- Many services are delivered without benefit of formal office space. These services may be delivered from stores, churches, restaurants, or vehicles. Gatekeepers, including mail carriers, beauticians, and neighbors, can be used for referrals. The cooperative extension network is recommended as a vehicle for educational programs. Aging, nutrition, and hospital programs can support and publicize a new endeavor. Programs that operate in isolation are less likely to be successful.

The second article, "Innovative Rural Mental Health Service Delivery for Rural Elders," was written by the same authors, listed in reverse order, and appeared on pp. 230-240 of the same issue of the same journal. Its findings were as follows:

- Distance, isolation, and resource shortages interfere with both problem identification and obtaining help with mental health issues. Ideally, an informal support system of family, neighbors, and friends helps rural elders through predictable life crises. Most services wait to be contacted, and traditional outreach tends to find those who are functioning fairly well. Several components common to successful direct service and educational programs are discussed.

- In many mental health programs, one individual's enthusiasm, work, and commitment were critical in organizing and persuading others to establish a program. A credible "natural leader" from the community knows how to present the concept in a way that would be acceptable locally and could motivate other groups to participate.

- In addition, services had to feel comfortable to clients. Impressions of comfort can be based on the program's appearance, location, time, and expense. Some programs provided services through nontraditional but trusted partners such as grain dealers, banks, and utility providers. Flexibility to choose to use portions of the services **when** they choose was important. Rural elder women tended to be cautious about accepting formal services and were hesitant to accept when they felt that they could not reciprocate.

- Two direct service models, gatekeepers and peer counseling, were described in this second article. Gatekeepers routinely have contact with people who themselves would not seek services. These gatekeepers make referrals to appropriate agencies. Peer counseling uses trained elder volunteers from the same community as part of the mental health team. Elders were more open to their own peers than they were to professionals, whom elders feared might threaten their independence. Peers often have a more realistic understanding of the client's situation.

- Three educational program models that trained nonprofessionals to recognize mental health problems and make appropriate referrals also were discussed. The nonprofessionals received cross-training about a number of organizations. The organizations had to gain a better understanding of the mental health needs of rural older adults and to collaborate in service provision. The three programs discussed were as follows:

  — A Missouri project trained trainers, using the well-established networks of area agencies on aging and university extensions. At eight sites in the State, local committees comprised of representatives of these two organizations, and community leaders, including clergy, community mental health providers, elder volunteers, or senior center directors, coordinated recruiting of candidate trainers who wanted training in mental health and aging. The project a) improved the providers' ability to recognize mental health needs, b) identified rural older adults who needed service, c) provided information on referral programs, d) assisted providers in notifying possible clients about their services, and e) helped providers inform agencies about possible clients.

— An Arizona program presented workshops for rural health (not mental health) staff, volunteers, family, and community members on identifying signs of mental illness in older adults, how to respond more effectively to people with these problems, and appropriate referrals. Culturally specific curricula were developed for rural Anglos, Latinos, and Native Americans. The local supports helped continue training after the program ended.

— A Pennsylvania program used a cross-system training model and focus groups to create a curriculum. Volunteers, gatekeepers, and nonprofessional caregivers were trained to give educational presentations in their communities. Recipients of training then conducted more training. This approach brought staff of the mental health and aging systems together at meetings.

- All three programs used committees and focus groups with community leaders in formal and informal roles. The committees helped develop the curricula specific to their locations and played key roles in marketing the program and recruiting people to participate. In addition, the importance of rural-specific material in such programs was stressed. Crisis intervention materials that advise calling 9-1-1 are not appropriate in areas that lack that service. Instead, the materials need to provide contact information for the appropriate local emergency contact, who may be the sheriff.

Index

D

E

www.ingramcontent.com/pod-product-compliance
Lightning Source LLC
Chambersburg PA
CBHW081442170526
45166CB00008B/2286